Understanding Fabless IC Technology

Understanding Fabless IC Technology

Jeorge S. Hurtarte
Evert A. Wolsheimer
Lisa M. Tafoya

AMSTERDAM • BOSTON • HEIDELBERG • LONDON • NEW YORK
OXFORD • PARIS • SAN DIEGO • SAN FRANCISCO
SINGAPORE • SYDNEY • TOKYO

Newnes is an imprint of Elsevier

Newnes is an imprint of Elsevier
30 Corporate Drive, Suite 400, Burlington, MA 01803, USA
Linacre House, Jordan Hill, Oxford OX2 8DP, UK

Recognizing the importance of preserving what has been written, Elsevier prints its books on acid-free paper whenever possible.

Library of Congress Cataloging-in-Publication Data
Hurtarte, Jeorge S.
Understanding fabless IC technology / Jeorge S. Hurtarte, Evert A. Wolsheimer, Lisa M. Tafoya.
p. cm.
Includes bibliographical references and index.
ISBN 978-0-7506-7944-2 (pbk. : alk. paper) 1. Integrated circuits—Design and construction.
2. Integrated circuits industry. I. Wolsheimer, Evert A. II. Tafoya, Lisa M. III. Title.
TK7874.H8835 2007
621.3815—dc22

2007023423

British Library Cataloguing-in-Publication Data
A catalogue record for this book is available from the British Library.

ISBN: 978-0-7506-7944-2

For information on all Newnes publications
visit our Web site at www.books.elsevier.com

07 08 09 10 11 12 10 9 8 7 6 5 4 3 2 1

Typeset by Charon Tec Ltd (A Macmillan Company), Chennai, India
www.charontec.com

Printed in the United States of America

Dedication

"Only Real Men Have Fabs."
Jerry Sanders III, CEO of AMD, c. 1991

"Only Real Men Go Fabless!"
FSA, c. 2007

This book is dedicated to our colleagues in the fabless semiconductor industry who, despite the loud criticism, broad skepticism and financial risk, had the courage and tenacity to design, manufacture and market their innovative products using the fabless business model. These mavericks knew in their hearts this was the way of the future.
Jeorge, Evert and Lisa

Contents

Acknowledgements

The authors would like to acknowledge FSA (formerly known as the Fabless Semiconductor Association) for its support in publishing *Understanding Fabless IC Technology*. This book uniquely embodies the collective wisdom of industry experts and represents the views of the entire fabless/outsourced semiconductor supply chain.

In 1994, Jodi Shelton, along with a half a dozen fabless CEOs, had the forethought to establish a semiconductor association to promote the fabless business model. At the time, this model was highly criticized as a temporary business model meant to take advantage of excess fab capacity while companies prepared to build fabs. Few believed this model would survive – much less become the dominant model for the industry. Ms. Shelton recognized this vision well before it blossomed. As proof, FSA has become the global voice for the entire fabless ecosystem with over 500 members and leadership representatives in five countries, while still under the current leadership of Ms. Shelton. Ms. Shelton has supported the efforts of this book as one of the many opportunities to share her vision of this industry.

We would also like to thank the early founders of the fabless model for sharing their experiences of establishing FSA: Jodi Shelton, Ted Malanzcuk, Art Mandell, Bill O'Meara, Dr. Robert Pepper and Art Turriff. We offer a special acknowledgement to Dr. Morris Chang who was the ultimate risk-taker in this industry, establishing the first dedicated, pure-play foundry, Taiwan Semiconductor Manufacturing Company Ltd. (TSMC).

We would also like to thank our FSA colleague Kristen Pillans, Director of Research, for her tireless efforts collecting research data and applying industry benchmarks to measure the success of the fabless model.

We thank the current FSA Board of Directors, under the Chairmanship of Dr. Sanjay Jha, chief operating officer and president, QUALCOMM CDMA Technologies Group and FSA's Technology Committee, under the leadership of Wayne K. Nesbit, senior vice president of operations at Mindspeed Technologies and former FSA Board of Director, for supporting this effort and encouraging FSA to dedicate resources documenting the success of this business model.

We also thank Dr. Richard Jaeger at Auburn University and Dr. Santanu Das, CEO of TranSwitch Corporation, for their advice and support during the development of this book.

Finally, the authors acknowledge the publisher, Elsevier Inc., notably Harry Helms for his desire to explain the fabless semiconductor business model and Michele Cronin and Chuck Glaser for shepherding *Understanding Fabless IC Technology* from editing to publication.

Jeorge, Evert and Lisa

Preface

Prior to the 1980s, the semiconductor industry was vertically integrated. Semiconductor companies owned and operated their own silicon wafer fabrication facilities and developed their own process technology for manufacturing their chips. These companies also performed the assembly and test for their chips.

Meanwhile, with the help of private-equity funding, smaller companies began to form, with experienced engineers exercising their entrepreneurial prowess by establishing their own IC design companies focused on innovative chip solutions.

As with most technology-intensive industries, the silicon manufacturing process was and is cost prohibitive, especially for these small start-up companies. These companies relied on using excess capacity from integrated device manufacturers (IDMs) to manufacture the chips they were designing.

This was the birth of the fabless business model. Companies were manufacturing integrated circuits (ICs) without a fab! Simultaneously, the foundry industry was established by Dr. Morris Chang with the founding of TSMC. The foundry industry would become the cornerstone of the fabless model – providing a non-competitive manufacturing partner for these innovative and pioneering fabless companies.

The fabless business model was sharply criticized as an opportunistic business "tool" soon after it was established in the 1980s. During the 1990s, however, industry pundits acknowledged the financial success of fabless companies, such as Nvidia, Broadcom and Xilinx. Today, it is the preferred business model for the semiconductor industry.

When FSA was established in 1994, there were only three companies – Cirrus Logic, Adaptec and Xilinx – with revenues in excess of $250 million. Today in 2007, FSA tracks 10 fabless companies that have surpassed $1 billion in annual revenues.

The model has been further validated by the conversion of major IDMs to a completely fabless model, including, for example, Conexant Systems, Semtech, and, most recently, LSI Logic. In truth, virtually all other IDMs have flirted with the model over the years, and, today most major IDMs including Freescale, Infineon, Texas Instruments and Cypress Semiconductor have adopted outsourcing as a major manufacturing strategy.

Since most semiconductor companies are either fabless or embracing the model to some extent, the need to thoroughly understand how to conduct business in the fabless, or outsourced, industry environment has increased. This is especially true for those employees within traditional IDM companies, who for many years have worked in a vertically integrated organization and now operate in a horizontal and outsourced industry environment. Likewise, new engineering and business graduates will find themselves working in a fabless environment during their career.

Fundamentally, the major difference between the IDM and fabless business models is resource control. With the IDM model, all fabrication facilities and related resources are under the direct control of a vertically integrated company. That is not the case, however, with the fabless model. The fabless business model is totally dependent on building complementary and lasting relationships with various suppliers. Today, the suppliers include not only manufacturing, but assembly and test, intellectual property (IP) and electronic design automation (EDA) companies, among others.

Understanding Fabless IC Technology intends to address these differences and provide the reader with an overview of how to conduct business in the fabless semiconductor industry environment as well as discuss technical, cultural and global issues.

This book addresses a number of questions:

- How does a fabless company transfer its design data to the foundry?
- How is work-in-process (WIP) information accessed from the foundry?
- What are the fabrication process design rules – how do fabless companies know when they are mature and ready for production?
- What about circuit SPICE models and design libraries – should they be obtained directly from the foundry or from third-party suppliers?
- Should assembly and test be performed by the same foundry supplier or by a separate company?
- Who handles product characterization?
- What roles do e-commerce and information technology (IT) and management information systems (MIS) play in the fabless model?

Once the appropriate fabless strategy has been evaluated and adopted, then there is the issue of implementing it in a global environment and adapting to different worldwide cultures. For example, the semiconductor design may occur in the United States, embedded software development may occur in India, the fabrication in Taiwan or China, semiconductor packaging and test may be handled in the Philippines, and final test may come back to the United States.

How do employees in a fabless environment deal with the various cultures, languages, time zone and social customs differences to ensure a smooth process from design to final product delivery to their customers?

The book is intended to support and be utilized by:

- Fabless employees who seek to gain a broader understanding of the overall industry;
- IDM employees transitioning their integrated environment to a fabless outsourced environment;
- Supplier employees who want an in-depth understanding of the issues and concerns faced by their current and potential customers;
- Investors and venture capitalists who require a thorough understanding of the business issues as they impact the valuation and funding of the fabless semiconductor environment;
- Legislative bodies and government organizations who need to understand and recognize the value of this business model as a significant source of innovation and research and development investment in high-tech arenas;
- Entrepreneurs planning a semiconductor start-up company;
- College students planning their careers in the fabless environment.

This book is endorsed by FSA and was fully engaged in the entire process for the production of this book.

Understanding Fabless IC Technology is the product of contributions: from more than 30 professionals at more than 15 leading companies from the fabless semiconductor industry; from articles previously published in *FSA Forum* (formerly named *Fabless Forum*); and is presented by FSA – the "voice" of the fabless business model.

The editors of *Understanding Fabless IC Technology* have more than 55 years of combined experience both at fabless and IDM companies, dealing with business and technical issues related to the design and manufacturing of semiconductor products.

PART 1

Manufacturing Strategies: Understanding Fabless IC Technology

CHAPTER 1

More than a Decade of Transition in the Semiconductor Industry

The fabless business model has transitioned from opportunistic tactics to a strategic imperative. The model began to emerge in the 1980s, as the scale necessary to create and build an integrated device manufacturer (IDM) business from the ground up became prohibitive from a cost and resource perspective. Excess capital spending in the early to mid-1980s largely catalyzed the initial development of the fabless business model, and excess capacity left over from capital mistakes made by IDMs. Industry legends such as Gordon Campbell of Chips & Technologies (acquired by Intel), Mike Hackworth of Cirrus Logic and now deceased Bernie Vonderschmidt of Xilinx used the excess semiconductor capacity created by many of the vertically integrated Japanese electronics companies to pioneer the fabless business model in the early to mid-1980s.

Wikipedia defines a fabless semiconductor company as one that "specializes in the design and sale of hardware devices implemented on semiconductor chips. It achieves an advantage by outsourcing the fabrication of the devices to a specialized semiconductor manufacturer called a semiconductor foundry or fab." The credit for pioneering the fabless concept is given to Bernie Vonderschmidt of Xilinx and Gordon A. Campbell, founder of Chips & Technologies. In fact, Chips & Technologies (C&T) was the first fabless semiconductor company.

1.1 FSA is Established

In 1994, 52 companies had the vision and foresight to create what has become an extremely important and robust association designed to promote, educate and drive the success of the fabless business model in the global semiconductor industry (Figure 1.1).

Today, FSA has more than 500 corporate members, and many smaller fabless companies are continuing to drive significant innovation and growth.

1.2 Early Success

Companies such as Chips & Technologies, Altera, Cirrus Logic and Xilinx emerged and were able to take full advantage of the excess capacity held by the vertically integrated Japanese

1	Actel [NASDAQ: ACTL]
2	Adaptec [NASDAQ: ADPT]
3	Advanced Hardware Architectures (now Comtech AHA Corporation)
4	Alex Brown & Sons (now Deutsche Bank Alex Brown)
5	Amkor [NASDAQ: AMKR]
6	AMI Semiconductor [NASDAQ: AMIS]
7	Aptos Semiconductor (acquired by ISSI)
8	ASPEC Technology (now Ingennus Corp.)
9	Aura Vision (acquired by Broadlogic)
10	Brooktree (acquired by Conexant, 1996)
11	Catalyst [NASDAQ: CATS]
12	C-Cube (acquired by LSI Logic, 2001)
13	Chartered Semiconductor (now Chartered Semiconductor Manufacturing)
14	Chips and Technologies (acquired by Intel, 1998)
15	Cirrus Logic [NASDAQ: CRUS]
16	Comdisco (out of business)
17	Compass Design Automation (out of business)
18	Cowen & Company (now SG Cowen Securities Corporation)
19	Cyrix Corporation (acquired by Via Technologies, 1999)
20	DSP Group [NASDAQ: DSPG]
21	Exar [NASDAQ: EXAR]
22	Hambrecht & Quist (acquired by Chase Manhattan Bank, 1999)
23	Genesis Microchip [NASDAQ: GNSS]
24	IMP, Inc. (acquired by Teamasia Semiconductors, 2000)
25	Information Storage Devices (now ISD, a wholly owned subsidiary of Winbond)
26	Integrated Circuit Systems (acquired by IDT, 2005)
27	Level One Communications (acquired by Intel (NCG fabless group))
28	Logic Devices [NASDAQ: LOGC]
29	Micro Linear (acquired by Sirenza Microdevices, 2006)
30	Mid-West Microelectronics (out of business)
31	NeoMagic [NASDAQ: NMGC]
32	NVIDIA [NASDAQ: NVDA]
33	Montgomery Securities (acquired by Nations Bank, 1997 (now Bank of America))
34	OPTi
35	Prudential Securities (now Prudential Financial)

Figure 1.1 FSA Charter Members

36	Paradigm (acquired by IXYS)
37	Quality Semiconductor (acquired by IDT, 1998)
38	QuickLogic Corporation [NASDAQ: QUIK]
39	Robertson Stephens (out of business, 2002)
40	S3, Inc. (sold graphics chip unit to Via Technologies, 2000; now S3 Graphics Co., Ltd.)
41	S-MOS Systems (merged with Epson (now Epson Electronics America))
42	Schlumberger (Semiconductor Test Division became NPTest and was acquired by Cadence, 2004)
43	Seeq (acquired by LSI Logic, 1999)
44	Sierra Semiconductor (now PMC-Sierra [NASDAQ: PMCS])
45	Silicon Architects (acquired by Synopsys, 1995)
46	Standard Microsystems [NASDAQ: SMSC]
47	Submicron Technology (out of business)
48	Taiwan Semiconductor Manufacturing Corp. (TSMC) [NYSE: TSM]
49	Tower Semiconductor [NASDAQ: TSEM]
50	Western Digital
51	Winbond [TSE: 2344.TW]
52	Worltek International (now QPL (US) Inc.)
Source: FSA	

Figure 1.1 (Continued)

technology companies. Through solid execution, good product planning and an aggressive acquisition strategy, Cirrus Logic drove the fabless business model extremely hard, and became the first fabless semiconductor company to surpass the billion-dollar annual revenue barrier in 1995 – all in 45 quarters!

Due to the success of the fabless pioneers, pure-play wafer foundries such as Taiwan Semiconductor Manufacturing Corporation (TSMC) and United Microelectronics Corporation (UMC) emerged to legitimize the fabless business model and create a long-term source of supply for dedicated design-only semiconductor companies. Given the wafer foundries' ability to aggressively develop process technology and put sufficient capacity in place, a new breed of fabless companies began to emerge that were able to take an outsourced manufacturing strategy as a given.

With their superior design capabilities, solid execution and a stable source of wafer supply, companies such as Broadcom and Nvidia were off and running. Broadcom became the next company to achieve billion-dollar revenues in 2000 – all in 36 quarters! Not to be outdone, Nvidia followed suit in 2001, achieving billion-dollar revenue growth the fastest – in 32 quarters (Figure 1.2)! Their record-fast revenue growth proved that creating multi billion-dollar fabless companies was readily achievable.

Company	Date	Quarters
Cirrus Logic	1995	45
Broadcom	2000	36
Nvidia	2001	32
Source: FSA		

Figure 1.2 Quickest Fabless Companies to Achieve $1 Billion in Revenue

Today, 10 fabless semiconductor companies in the world report annual revenues in excess of $1 billion (Figure 1.3) and 69 more with annual revenues exceeding $100 million – and these numbers are growing each year – see Appendix showing all public semiconductor companies listed by 2006 revenue ranking.

Rank	Company Exchange	Stock Symbol	Ticker	CY2006 Revenue ($000)
1	QUALCOMM (QCT Division)	NASDAQ	QCOM	$4,331,000
2	Broadcom	NASDAQ	BRCM	$3,667,818
3	SanDisk Corporation	NASDAQ	SNDK	$3,257,525
4	NVIDIA Corporation	NASDAQ	NVDA	$3,068,771
5	Marvell Technology Group Ltd.	NASDAQ	MRVL	$2,237,596
6	LSI Logic	NYSE	LSI	$1,982,148
7	Xilinx, Inc.	NASDAQ	XLNX	$1,871,604
8	MediaTek Incorporated	TSEC	2454	$1,624,486
9	Avago Technologies	Private	Private	$1,576,000
10	Altera	NASDAQ	ALTR	$1,285,535
Source: FSA				

Figure 1.3 "Billion-Dollar Club" Companies

1.3 Early Success Trend Continues

When FSA was founded, fabless semiconductor companies recorded about $3 billion in revenues. As of 2006, that revenue figure had grown to $49.6 billion (Figure 1.4).

Although fabless revenues only account for 20 percent of all semiconductor revenue, the fabless growth rate is phenomenal. For example, while the overall semiconductor industry has grown more than 140 percent since 1994, the fabless sector has increased more than 1,400 percent!

The fabless growth rate will remain cyclical, trending with the overall semiconductor industry. But despite the cyclicality inherent in the overall semiconductor industry and fabless

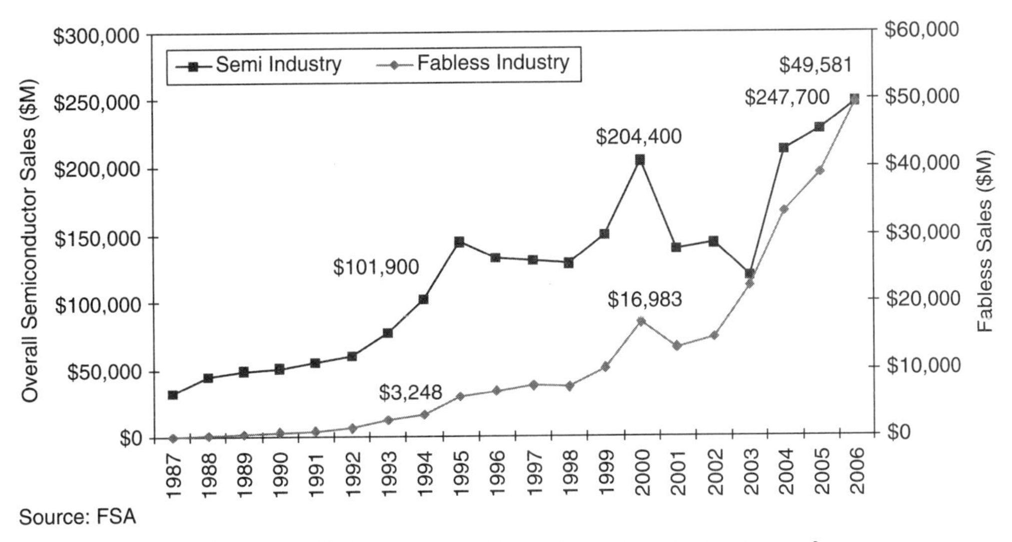

Figure 1.4 Fabless versus Overall Semiconductor Growth

sector, FSA forecasts that fabless revenues will continue to grow faster than the overall semiconductor industry during the next decade.

1.4 Semiconductor Business Models

Depending on their manufacturing strategy, semiconductor product companies can be segmented into two categories:

- *Integrated Device Manufacture*. IDMs have their own semiconductor manufacturing plants and resources and thus have direct control over their capacity, processes and resources. Examples include Intel, Samsung and most Japanese semiconductor companies.
- *Fabless*. These companies outsource all semiconductor manufacturing to foundries like TSMC, UMC, Chartered Semiconductor and SMIC. Examples of fabless companies are: Broadcom, Nvidia, Xilinx and TranSwitch.
- *Hybrids*. These are typically IDMs who outsource some of their manufacturing requirements to foundries when they need some additional capacity or some other processes not available in-house. This hybrid strategy allows the IDMs to bypass or delay additional capital expenditures.

In addition to the semiconductor product companies, the *foundries* make their business model that of performing 100% semiconductor manufacturing services for all of above product companies. As such, foundries do not design or market products under their brand name, and thus are not viewed as competitors by their customers.

1.4.1 Integrated Device Manufacturer

Companies like Intel and Samsung have traditionally been categorized as Integrated Device Manufacturers or IDMs. The IDMs have total ownership and control of all the required resources for semiconductor manufacturing: design automation tools, process libraries, process technology, fabrication and assembly equipment, test equipment, and all the necessary facilities and infrastructure resources. Thus, IDMs need to constantly be preoccupied with capital expenditures for their increasing production demands, technology changes, and under-capacity utilization costs in times of declining demand.

It can be said that every minute that an IDM CEO spends in discussing the company's internal manufacturing capital expenditures requirements, is a minute *not* spent discussing the next product generation roadmap or a minute *not* spent with a prospective customer, thus in a way, taxing critical resources throughout the company. On the other hand, some IDMs have pursued this manufacturing strategy due to the belief that they can differentiate themselves in terms of a leading edge semiconductor fabrication or packaging process or technology. This differentiation is becoming less of a reason as the leading foundries are catching up in bringing leading edge technology at par with the IDMs.

1.4.2 Fabless

Unlike IDMs, fabless semiconductor design companies do not own any manufacturing facilities and thus are totally focused in the design, marketing and sales of their products. Fabless companies are not "distracted" by the worries of intensive manufacturing capital investments and the associated internal resources required to keep the manufacturing facilities in operation. Fabless companies are at freedom to select any of the foundries that fit their needs in terms of technology, price, service or location for example. This selection freedom allows them to negotiate the best value to meet their needs and thus they are not "tied" down to a single supplier like in the case of the IDMs.

Figure 1.5 shows the top 20 fabless vendors by 2006 revenue. Total fabless revenue was nearly $50 billion in 2006, with a 12-year CAGR of 25 percent, compared to 8 percent for IDMs.

Rank	Company Exchange	Stock Symbol	Ticker	CY2006 Revenue ($000)
1	QUALCOMM (QCT Division)	NASDAQ	QCOM	$4,331,000
2	Broadcom	NASDAQ	BRCM	$3,667,818
3	SanDisk Corporation	NASDAQ	SNDK	$3,257,525
4	NVIDIA Corporation	NASDAQ	NVDA	$3,068,771
5	Marvell Technology Group Ltd.	NASDAQ	MRVL	$2,237,596
6	LSI Logic	NYSE	LSI	$1,982,148

Figure 1.5 Top 20 Fabless Companies by 2006 Revenues

Rank	Company Exchange	Stock Symbol	Ticker	CY2006 Revenue ($000)
7	Xilinx, Inc.	NASDAQ	XLNX	$1,871,604
8	MediaTek Incorporated	TSEC	2454	$1,624,486
9	Avago Technologies	Private	Private	$1,576,000
10	Altera	NASDAQ	ALTR	$1,285,535
11	Conexant Systems	NASDAQ	CNXT	$985,615
12	NovaTek	TSEC	3034	$964,314
13	Himax Technologies	NASDAQ	HIMX	$744,518
14	Cambridge Silicon Radio (CSR)	LSE	CSR.L	$704,700
15	VIA Technologies, Inc.	TSEC	2388	$657,901
16	QLogic Corporation	NASDAQ	QLGC	$570,051
17	OmniVision Technologies, Inc.	NASDAQ	OVTI	$540,741
18	Zoran Corporation	NASDAQ	ZRAN	$495,846
19	Silicon Laboratories	NASDAQ	SLAB	$464,597
20	Silicon Storage Technology, Inc. (SST)	NASDAQ	SSTI	$452,092
	2006 Top 20 Fabless - Total Revenue			**$24,902,483**
	Percent of Total Fabless Revenue 2006			50.21%
Source: FSA				

Figure 1.5 (Continued)

1.4.3 Hybrids

Hybrid manufacturing strategies are primarily used by IDMs for any or all of the following reasons:

- Complement their internal capacity constraints in times of up demand cycles, thus avoiding or delaying additional capital expenditures.
- As a strategy for benchmarking their internal operations and cost performance against the foundries.
- To gain access to manufacturing technologies not yet developed in house or with no plans to be developed in house.
- To get additional revenues off of their excess capacity in down demand cycles while still retaining their IDM model.
- To differentiate themselves with their own innovative semiconductor fabrication or packaging processes. However, with logic processes fast becoming commodities, the real differentiator is in the integrated circuit (IC) design itself, not the manufacturing.

Figure 1.6 A Look Inside a Fab (Courtesy of UMC)

In spite of all of the above reasons, even the IDMs at the top of the chip market rankings are gradually relinquishing control of manufacturing. Foundry process technology is so advanced that IDMs recognize process technology can be bought instead of being internally developed today.

1.4.4 Foundries

Semiconductor foundries are companies that dedicate 100% of their business model to the manufacturing of chips for fabless IC semiconductor companies. Thus, foundries are not engaged on designing or marketing any of their own products and thus are viewed as a "neutral" manufacturing company which is not competing with their various product IC customers.

To stay in business, foundries like TSMC and UMC in Taiwan, Charter Semiconductor in Singapore, and SMIC in China, need to continuously keep investing on capital intensive semiconductor fabrication (thus their name "fab") equipment and related facilities and support infrastructure resources. This investment is required to keep up with capacity requirements, technology node roadmap progression, wafer size changes, and new state-of-the-art fabrication equipment.

Figure 1.6 shows the inside of a factory, where workers wear the typical "bunny suit" required inside the "clean rooms" to avoid any particle contamination.

1.5 Outsourcing Will Accelerate

Outsourcing trends will accelerate. Today, there are hundreds of fabless semiconductor companies, and new ones are being founded all the time. To support their growth, leading-edge wafer foundries have continued to develop and hone their skills. Going forward in today's complex and costly semiconductor environment, it is hard to imagine that a new semiconductor company could be successfully founded as a manufacturing company. The capital costs are simply too high. In comparison, the fabless business model has continued to expand, develop and prosper. As the manufacturing needs of fabless semiconductor companies have continued to increase, the wafer foundry community has continued to expand.

Chartered Semiconductor and SMIC have both emerged as large suppliers that are capable of generating yields that rival their veteran competitors and IDMs, and FSA believes that additional wafer foundries will become large suppliers as well. In 2006, from a revenue perspective, TSMC and UMC currently rank among the top 6 and top 24 largest semiconductor manufacturing companies in the world, respectively, according to Semiconductor Insights (Figure 1.7).

When FSA was founded in 1994, the semiconductor industry enjoyed 5- and 10-year CAGRs that were in the mid-to-high teens, with the industry exceeding $100 billion in annual revenues for the first time. Demand was explosive and was being driven by higher semiconductor content in PCs to support Microsoft's Windows operating system and the early ramp of analog cell phones.

2006 Rank	Company	CY Revenue ($000)	Headquarters
1	Intel	$31,580,000	US
2	Samsung Electronics	$19,670,000	South Korea
3	Texas Instruments	$13,730,000	US
4	Toshiba	$10,030,000	Japan
5	STMicroelectronics	$9,855,000	Europe
6	TSMC	$9,759,000	Taiwan
7	Renesas	$8,170,000	Japan
8	HYNIX	$7,375,000	South Korea
9	Freescale	$6,080,000	US
10	NXP	$5,935,000	Europe
11	NEC	$5,725,000	Japan
12	Micron Tech	$5,510,000	US
13	AMD	$5,245,000	US
14	Infineon	$5,055,000	Europe
15	Qimonda	$5,005,000	US
16	Sony	$4,742,000	Japan
17	Matsushita Electric Industrial	$4,450,000	Japan
18	IBM Microelectronics	$4,400,000	US
19	QUALCOMM (QCT Division)	$4,331,000	US
20	Broadcom	$3,667,818	US
21	Sharp	$3,490,000	Japan
22	Elpida	$3,450,000	Japan
23	SanDisk Corporation	$3,257,525	US
24	UMC	$3,196,000	Taiwan
25	Fujitsu	$3,116,000	Japan
Source: FSA, Semiconductor Insights			

Figure 1.7 Top 25 Semiconductor Sales Leaders

Following 29 percent growth in 1993, the semiconductor industry grew 32 percent in 1994 and onto 42 percent growth in 1995. However, since then, several downturns have affected industry growth, and over the past couple years, industry growth has slowed to 7 percent in 2005, up to 9 percent in 2006 (Figure 1.8).

Today's growth is driven heavily by communications (wireless and wireline) and consumer electronics, with PC applications still being a significant element of the semiconductor market (Figure 1.9).

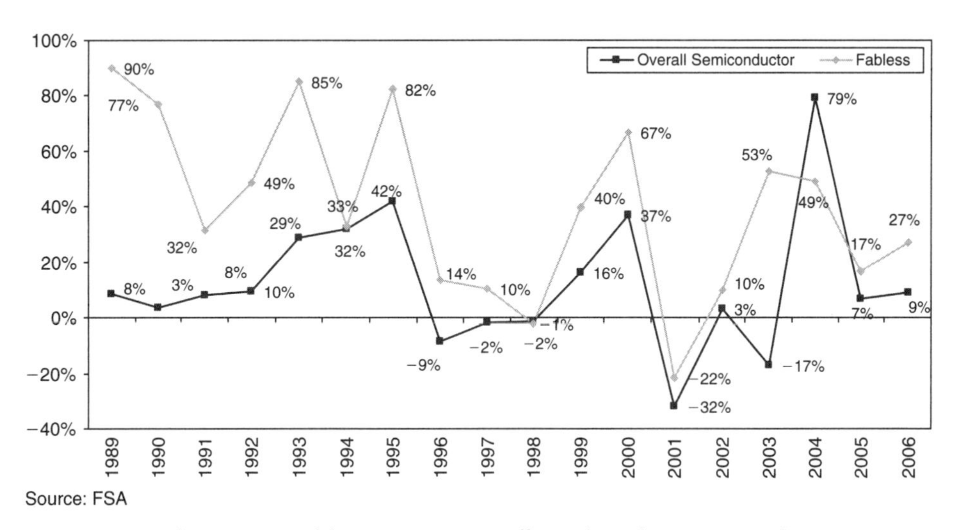

Figure 1.8 Fabless versus Overall Semiconductor Growth

	2006 ($M)	%	2007 ($M)	%	2011 ($M)	%	2006–2011 CAGR (%)
Automotive	20,043	8	22,444	8	29,442	7	8
Computer	89,082	36	95,800	35	142,603	35	10
Consumer	53,497	22	61,356	22	91,472	23	11
Wired	23,140	9	25,096	9	33,838	8	8
Wireless	39,559	16	45,965	17	75,587	19	14
Industrial	21,899	9	25,419	9	29,890	7	6
Total	**$247,220**	**100**	**$276,079**	**100**	**$402,832**	**100**	**10**
Source: Databeans, 2007							

Figure 1.9 Semiconductor Application Breakdown and Growth

In the midst of the strong cyclical expansion phase in the 1994 timeframe, the ability to secure capacity was one of the most important issues of the day, and FSA served as a non-competitive conduit to support the legitimacy of foundries and the fabless semiconductor sector's way of life. It is clear that FSA's founders had excellent vision when they created this organization to support the needs and initiatives of fabless companies.

In 1994, semiconductor companies were making equity investments in foundries to ensure capacity in what was then a very supply-constrained environment. While high levels of excess capacity caused by sluggish demand and significant capacity overbuild in the mid-1990s rendered many of these investment strategies unnecessary, FSA's efforts clearly helped to raise the awareness that semiconductor design companies that could look to foundries to support their manufacturing needs. In addition to legitimizing the fabless business model, FSA has enhanced the overall level of intelligence within the semiconductor industry's supply chain.

FSA encourages fabless companies to continue to innovate and gain share. Due to the law of large numbers and the challenges with meaningfully increasing the semiconductor content in electronics from the high content in many of today's products, FSA expects the long-term growth rate of the semiconductor industry to continue its single-digit growth. However, fabless companies should continue to enjoy strong relative performance when compared to the overall industry. Furthermore, the need for strong support from foundries is imperative to enable the long-term growth potential of the fabless companies and the overall semiconductor industry.

Due to their ability to focus all of their research and development (R&D) efforts on design, rather than expend efforts designing process technologies and ramping manufacturing capacity as well, the fabless company segment has been responsible for many of the key innovations in the overall semiconductor industry. For example, Xilinx invented field programmable gate arrays (FPGAs), Silicon Laboratories pioneered CMOS transceivers and power amplifiers for cellular phones and Nvidia drove the revolution in 3D graphics for PCs.

In addition to pioneering new technological innovations, many fabless companies have emerged in the last one to two decades to become some of the fastest-growing and most vibrant companies in the overall semiconductor industry, including Altera, Broadcom, Marvell, Nvidia, Silicon Laboratories and Xilinx.

Due to the extreme capital intensity of the overall semiconductor industry and fast pace of advancing technology, this industry will continue to be characterized as volatile and cyclical, and FSA will enhance the long-term direction of its member companies and the future leaders within the industry.

1.6 IDMs are Going Fabless

1.6.1 Semiconductor Firms are Forging a New Path

Fifteen years ago, it was unthinkable to start a semiconductor business without resources significant enough to build a fab. The emergent company needed to possess all of the competencies required to bring its technology to market, from innovative concepts in architecture and circuit design through process chemistry and procedure; to marketing vision, distribution and access. The fabless model emerged to challenge this paradigm.

Unprecedented technological complexities and the economics of implementation drive the pace of change within the semiconductor industry. Not only are fabless IC suppliers proving the viability of an outsourcing business model in the global semiconductor industry, they are the harbingers of change.

IDMs are benefiting from the infrastructure in place to provide service to the industry. Devices are fabricated at pure-play foundries, and packaging and test houses perform value-added services for the fabless and integrated companies alike.

The cost of developing and deploying new technology is escalating. Consider today's realities: advanced processes are integrating in excess of 100 million transistors on a chip, and pundits are predicting one billion in the next decade (Figure 1.10).

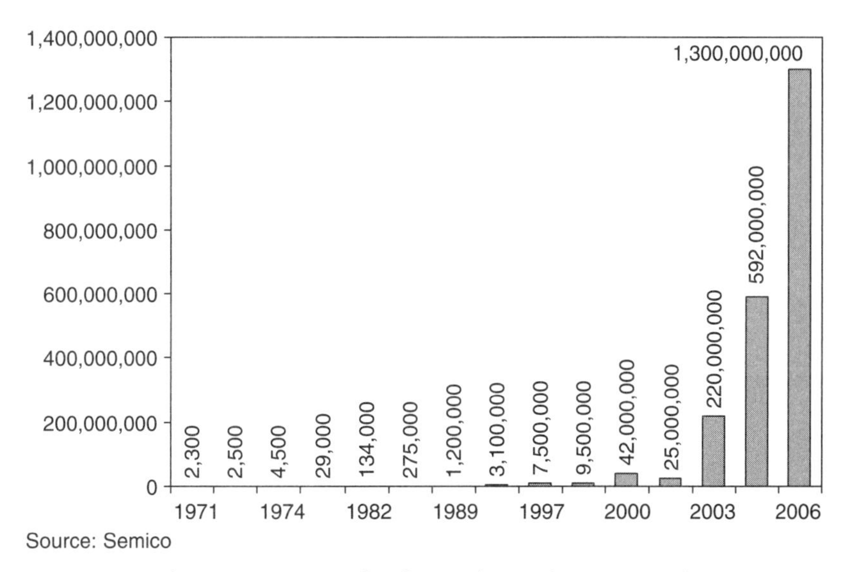

Figure 1.10 Growth of Transistors (1971–2006)

The process shrink from 0.80-micron to 0.25-micron, 0.18-micron, 0.13-micron, to 90-nanometer and 65-nanometer has occurred as forecasted, and the pace is accelerating as the industry moves along a roadmap taking it toward 45-nanometer and 32-nanometer (Figure 1.11). Simultaneously, the burgeoning market has driven average selling prices (ASPs) for new technology devices down at unprecedented rates.

Among the variety of methods companies are using to secure resources and minimize risk, the most prominent is the strategy of outsourcing.

1.6.1.1 Outsourcing: A Fundamental Model for Success

"Core competency" is a prescript that has rapidly permeated every corner of the electronics industry.

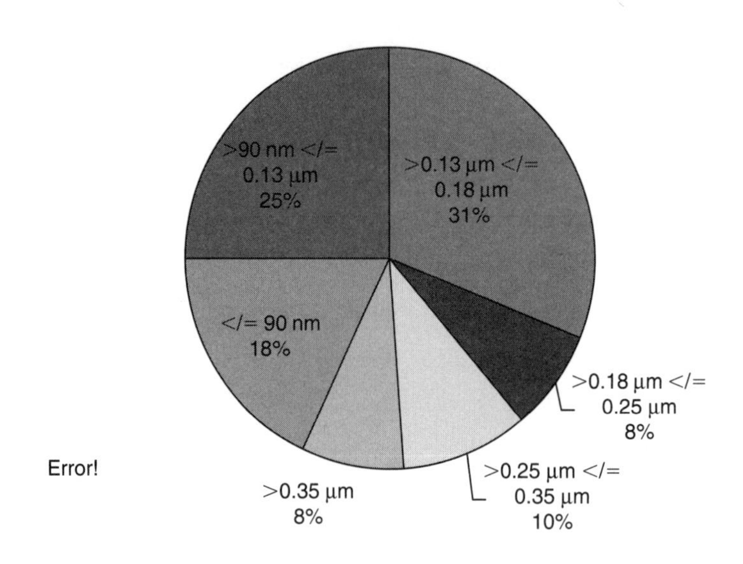

Source: Company Reports, IC Insights

Figure 1.11 The 2006 Pure-Play Foundry Sales by Feature Size

Within the electronics industry, this business tactic was implemented by the electronics original equipment manufacturer (OEM) because it became too economically painful, and at times impossible, to remain vertically integrated. It is an extremely rare electronics manufacturer that does not outsource some aspect of its design, manufacturing, packaging or shipping functions. This trend is reflected in the steady growth of contract manufacturers. This has been fueled not only by continued outsourcing, but also by the vertical integration of services, from design through build and on to final delivery.

This has now spread to supplier firms as well as to semiconductor firms in particular. The fabless semiconductor business framework is a case in point. The fabless firm is a company that has value to add, but cannot or will not allocate resources to fab construction and operation. Instead, that aspect of the business is outsourced. Innovative firms are taking this successful business model and extending it one step further by outsourcing design, as well.

In each instance, outsourcing is a tactic that companies use to free themselves of activities that others can do more efficiently. The company that outsources does so to reduce risk, reduce costs, enhance quality, enhance market access and improve profitability. Outsourcing has proven itself as a fundamentally sound business practice, and the core concept has been replicated throughout the electronics industry.

Companies must balance the technical, managerial and financial challenges that are distracting semiconductor companies from their core competency.

Capital utilization in the semiconductor industry stands at about 80 percent in many operations (Figure 1.12). Companies that are not involved in raising their capital utilization to 70 percent and beyond risk becoming non-competitive. Whether it is making refinements on front-end processes or substantial improvements in back-end efficiencies, every semiconductor manufacturer must make it happen. Each will decide whether to allocate its resources to this challenge, or to outsource these technical and managerial challenges to a competent resource partner.

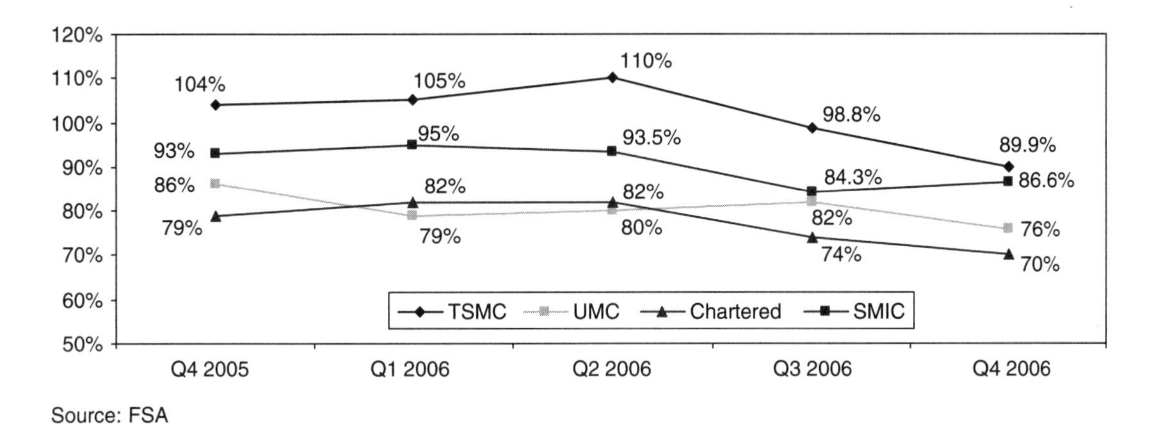

Source: FSA

Figure 1.12 Top 4 Pure-Play Foundry Utilization Rates (Q4'05–Q4'06)

As integration levels include advanced systems-on-a-chip (SOC) design, system-in-package (SiP) design with embedded SOCs and million of transistor counts, many operations are experiencing an extraordinary spiral of demand for technical and managerial attention. The need for design-for-test (DFT) strategies and design for manufacturability (DFM) are two examples. Each semiconductor firm must decide whether to expend its own resources to resolve these issues, or to outsource the solution while concentrating on its own areas of distinctive competence, or to opt for some blend of both.

As functionally integrated circuitry becomes more complex, the challenges of silicon debug and repair are growing proportionally. Shrinking product life cycles impact the critical need to speed silicon debug in a race to market. This is particularly acute in the areas of communications and consumer electronic applications.

Each step in the semiconductor process – from concept through design, into the diverse and multiple manufacturing steps, through test, packaging, warehousing and shipping – is a transaction. And each transaction can be handled internally or outsourced to a competent partner. Further, each transaction has both a technical and a managerial component which can be outsourced independently or bundled.

1.7 A Case Study: Cypress Semiconductor

Historically, Cypress Semiconductor long maintained that owning its own fabrication facilities was the only way for a large, diversified semiconductor company to compete – and win – in demanding, high-volume markets. So many were surprised when Cypress selected TSMC to produce two new families of complex programmable logic devices (CPLDs).

But the decision to "go fabless" in programmable logic devices (PLD) made sense for Cypress for several reasons – not least of which was its belief that access to advanced pure-play manufacturing technology would enable Cypress to produce the biggest, fastest, most cost-effective CPLD family on the market much more quickly than it could do itself.

Cypress has long been in the PLD business – in fact, PLD was one of its original product divisions when it went public in 1986, and it partnered with Altera to produce the original MAX family of CPLDs. Cypress maintained a leadership position for many years; however, in the early 1990s, Cypress began to fall behind in PLD process technology. Part of the reason was that the Static Random Access Memory (SRAM) market became more competitive, forcing it to concentrate its development resources more heavily in that area. Cypress' business model was to develop the next SRAM process, perfect it, and only then build the derivative processes required for other products, including PLDs. As a result, Cypress fell behind in the advanced technology capability. At the same time, its engineers developed the architecture and software for the two families of PLDs that could make Cypress a leader in the programmable logic market. So, it was faced with a critical choice: develop and fab the new families itself, or expand the traditional Cypress business model and seek an outside manufacturing partner.

Cypress knew that there would be several advantages to developing and manufacturing the line itself. For a time, that choice seemed obvious. Cypress believed it would have increased control and more flexibility. For example, if something went awry in the development process, Cypress felt it could fix the problem immediately, and, it believed it could craft the process to provide exactly the features it desired.

But then reality set in. Cypress began to recognize that attempting to make the PLDs it sought to produce with its in-house SRAM processes created mismatch after mismatch in terms of feature sets. Cypress' "SRAM-first" development hierarchy also meant it was always behind the curve in terms of density and performance. After adding additional metal layers to an SRAM process to handle a PLD's interconnect requirements, creating new design rules and adapting libraries to meet the needs of a new device, Cypress found the process it needed always significantly lagged the equivalent line width SRAM process, sometimes by as much as a year and a half.

These problems compounded those it had expected: that for Cypress to manufacture the chips itself would be costly and would divert precious engineering resources away from other projects. Cypress management recognized that it needed to rapidly develop multiple technologies, including non-volatile technology and embedded RAM, at different lithographies with both 5 volt and 3.3 volt capability. And this was in addition to the needed investment of tens of millions of dollars for architecture, software, design and customer support.

So, Cypress began to look more closely at another option: developing a foundry partnership. The advantage was hard to deny: using a foundry would mean a low-up-front investment (in exchange, of course, for Cypress' commitment to purchasc wafcrs). Instant access to the same variety of high-end technologies as the other major competitors in its business meant it would be able to focus its R&D dollars on architecture, software and features.

But the apparent disadvantages of a foundry partnership still remained. Using a foundry would prevent Cypress from differentiating its product families on the basis of process technology. And, not paying for process R&D up-front just meant it would pay later in the form of reduced gross margin. Finally, Cypress' internal manufacturing organization and systems were designed to work with fabs that were part of the corporate umbrella, not independent entities.

Ultimately, Cypress rationalized these disadvantages in several ways. First, in the PLD business, state-of-the-art process technology is a requirement to be in the business – think of it as the price of entry. Cypress has long been cognizant of the fact that architecture, software and service are the real value-adds for PLDs. Being fabless just means the stakes are even higher in these three areas, since highly specialized process technology and direct wafer manufacturing would be left behind.

Developing and integrating a foundry-based manufacturing flow and systems into existing Cypress structures was mostly a matter of adapting its existing systems. Cypress just needed to form the team and go do it.

When selecting potential foundry partners, Cypress considered all the major foundries. Ultimately, Cypress selected TSMC because it was the world's largest dedicated IC foundry and it offered a comprehensive set of IC fabrication processes, including those to manufacture CMOS logic; mixed-mode, volatile and non-volatile memory; and BiCMOS chips. In addition, Cypress management felt TSMC provided the company with the best cultural fit.

In February 2007, Cypress Semiconductor announced it plans to more than double its production outsourcing ratio over the next 5 years. UMC is Cypress' foundry partner on the production of 65-nanometer SRAM. Cypress indicated UMC will be the major foundry for 65-nanometer and below processes.

Cypress plans to increase its production proportion from the present 20 to 50 percent over the next 5 years. Cypress already has many foundry partnerships with several pure-play foundries including China-based Grace Semiconductor Manufacturing Corporation (GSMC) and Korea-based Megachip, and it has indicated it will maintain having multiple foundry partners in the future.

Foundries want to partner with fabless and IDMs alike. In Cypress' case, TSMC established itself as an important partner and captured a share of the wafers that Cypress would otherwise have built itself.

Cypress also maintains a stable outsourcing volume at TSMC each quarter, with no stated plans to discontinue that relationship.

1.8 More IDMs are Outsourcing

In early 2007, TI made a significant change when it announced that it will complete the development of its own 45-nanometer logic process. Following that, it decided to stop internal development at the 45-nanometer node and use foundry-supplied processes at 32-nanometer, 22-nanometer and thereafter. TI develops a process and aligns the technology with various foundry providers, mainly Chartered, UMC and TSMC. This allows the company to avoid duplication and reduce its research and development (R&D) costs.

For the high-performance computing/microprocessor arena, TI manufactures Sun Microsystems Inc.'s Sparc-based processors on a foundry basis for the workstation company. For this segment, TI develops a high-end process technology and also makes the processors within its own fabs, as does Fujitsu Ltd. Also, over time, TI will work with perhaps one foundry on the joint development of a process technology for Sparc-based processors. That foundry may also manufacture the chips. Going forward, it has indicated that it will probably work with only one foundry partner on microprocessor class products/Sun products – one that is the closest to a "joint development" model. The high-performance DSP/ASIC (digital signal processing/application-specific integrated circuits) will come directly from the foundry with reduced involvement by TI.

TI has not indicated it will change its analog strategy. But going forward, it will be working closely with foundries to jointly define low-power processes and work with multiple foundries. It is likely that TI will keep its current foundry partners in the low power and DSP/ASIC areas where TI works with Chartered, UMC and TSMC. And in analog, the chip maker also develops and makes those products within its own plants.

The long-standing Crolles2 Alliance (a joint venture between IDMs: ST, Freescale and NXP (formerly Philips Semiconductor) for joint process development at advanced nodes disintegrated in early 2007. Former members Freescale, ST and NXP all separately announced plans to move in different directions and form alliances with other fabs or foundries.

Freescale has joined the Common Platform (Chartered, IBM and Samsung) for development of 45-nanometer and below.

ST plans to shift gears and align with unnamed industry leaders at the 32-nanometer node.

NXP announced in April 2007 that it would outsource all of chip production beyond 90-nanometer to TSMC. NXP is investing around $1.34 billion annually in R&D. As a result, it decided to redirect its investment to focus on differentiation on top of the TSMC platform rather than trying to reinvent a platform itself, given that the technology that it was developing in Crolles is available from TSMC. Attempting to develop the technology itself would only duplicate expensive investment and likely cause NXP to be late to market with its chips.

With the cost of 300-millimeter (12-inch) wafer fabrication facilities approaching $5 billion or more, coupled with the soaring price of process development and IC equipment, the decision to work more closely with foundries is a necessary choice if IDMs are to control their costs while keeping up with Moore's Law. Figure 1.13 shows TSMC's 2007 estimate of how the cost of building fabs at various levels of production is escalating. Figure 1.14 demonstrates UMC's historical growth of fab development costs.

Increasingly, IDMs are allowing the foundries to handle not only more of the traditional IC production requirements, but also the R&D. Over time, process technology will become less of a differentiator for many chip companies, at least on the digital CMOS side. IDMs will

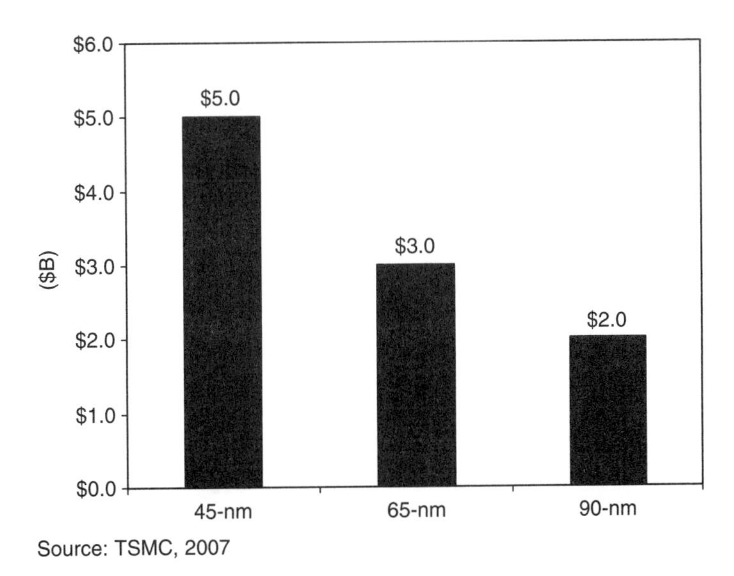

Figure 1.13 Fab Cost is Escalating by the Process Node

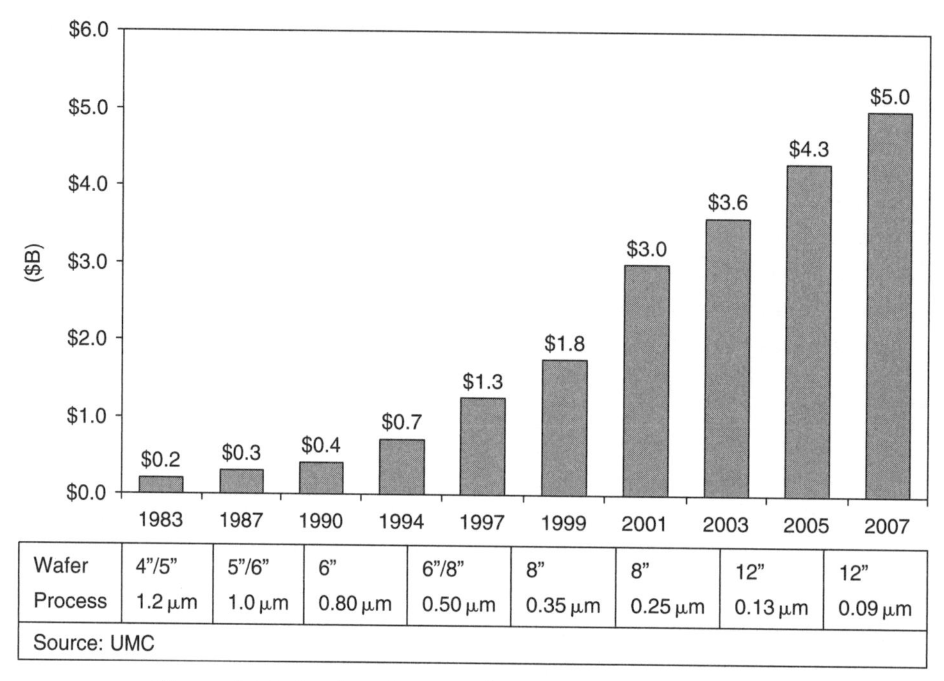

Wafer Process	4"/5" 1.2 μm	5"/6" 1.0 μm	6" 0.80 μm	6"/8" 0.50 μm	8" 0.35 μm	8" 0.25 μm	12" 0.13 μm	12" 0.09 μm
Source: UMC								

Figure 1.14 Semiconductor Fab Cost Trend (1983–2007)

have little or no process technology advantage over their fabless rivals, many of which use the same foundries to gain access to cutting-edge technologies.

The key to the outsourcing strategy for IDMs is collaboration. Fabless companies have long understood the concept of foundry "partner" and now IDMs are recognizing the value of collaboration to share the R&D expenses and build process technology with other advanced node leaders.

1.9 Geographic Manufacturing Centers

This section provides a general overview of the most prominent semiconductor manufacturing locations for both IDMs, which have their own internal manufacturing facilities, and foundries.

As shown in Figure 1.15, most IDM manufacturing locations are in Japan and North America. Other locations include Korea, Europe and Israel and to a lesser degree, Taiwan, Singapore and China. In sharp contrast, most of the foundry locations are in Taiwan, Singapore and China, and to a much lesser degree in the rest of the other countries. Some foundries are

	Japan	North America	Europe and Israel	Taiwan	Korea	China	Singapore
Key IDMs	Renesas	Intel	Infineon	Winbond	Samsung	Huawei	DenseLight Semiconductors
	Toshiba	Texas Instruments	STMicro	Nanya	Hynix		
	NEC	Freescale	NXP	Powerchip	MagnaChip	Huajing Micro	
	Sharp	AMD	AMD	ProMOS			
	Matsushita	Micron	Intel			Lenovo	
	Sony	Analog Devices	Robert Bosch				
		IBM					
Major Foundries	Toshiba	IBM Micro	X-Fab	TSMC	Hynix	SMIC	Chartered
	Seiko-Epson	Jazz Semiconductor	Tower Semi	UMC	DongbuAnam	ASMC	
	Fujitsu					Grace	
Key Products	Microcontrollers	MPUs	Logic	Logic	Memory	Memory	Memory
	Discretes	MCUs	ASSPs	ASSPs	Logic	Logic	Analog
	Memory	DSPs	Analog	Memory	ASSPs	ASSPs	Logic
	Logic	Logic	Discretes	Display Drivers			
	ASSP	ASSPs	Memory				
	Analog	Analog	MEMS				
		Memory					
		Discretes					

Source: FSA

Figure 1.15 Key IDMs, Foundries and Products by Region

spin-offs of previous IDMs (for example, Jazz Semiconductors in the U.S. was a spin-off of Conexant Systems when it transitioned to fabless).

Of interest is also the product specialization trend across the various manufacturing regions. In general, it can be seen a greater degree of specialization in memory and general purpose logic chips in Asia, and more specialization in MPU/MCU/DSPs in North America.

CHAPTER 2

Fabless Semiconductor Manufacturing

As the fabless business model becomes the preferred model for semiconductor product companies, the foundries are not only keeping pace with the manufacturing capacity, pricing and service needs of their fabless customers, but also are starting to lead in adopting the next process technology node at par with their IDM competitors (some of which are also their customers). Figure 2.1 shows the timing for the adoption of the 45-nanometer technology node for various leading foundries.

Foundry	Timing
TSMC	Sept. 2007
UMC	Q4 2007
Chartered	Q4 2007
IBM	Q4 2007
Samsung	Q4 2007
Source: TSMC, UMC, IBM, April 2007.	

Figure 2.1 45-nanometer Foundry Release Schedule

2.1 Foundry Revenue Growth

Foundry revenue growth has consistently outpaced the growth of the semiconductor industry and is a good measure of the success of the fabless business model.

Figure 2.2 shows the foundry revenue and market share. TSMC leads the pack with a 2006 market share of 45.9 percent, followed by UMC with 14.7 percent, and then by SMIC with 6.7 percent. Of interest is the rapid ramp up of SMIC, based in China, to claim the third place in 2006 ahead of Chartered Semiconductor.

Another metric for foundry revenue and market share is revenue per wafer produced. Figure 2.3 shows this metric for the top four foundries. By this metric, TSMC still leads with a 49 percent market share, followed by UMC at 19 percent, and then by both Chartered and SMIC at 7 percent.

	2000		2002		2004		2006	
	$M	%	$M	%	$M	%	$M	%
TSMC	$5,325	42.3	$4,654	46.6	$7,653	42.6	$9,875	45.9
UMC	$3,183	25.3	$1,939	19.4	$3,670	20.4	$3,168	14.7
SMIC	$0	0.0	$50	0.5	$975	5.4	$1,439	6.7
Chartered	$1,134	9.0	$449	4.5	$932	5.2	$1,405	6.5
IBM	$392	3.1	$760	7.6	$731	4.1	$698	3.2
Samsung	$26	0.2	$71	0.7	$196	1.1	$526	2.4
Powerchip	$108	0.9	$63	0.6	$331	1.8	$478	2.2
DongbuAnam	$354	2.8	$248	2.5	$382	2.1	$419	1.9
Vanguard	$584	4.6	$240	2.4	$499	2.8	$398	1.8
MagnaChip	$298	2.4	$210	2.1	$262	1.5	$324	1.5
HHNEC	$142	1.1	$145	1.5	$227	1.3	$320	1.5
X-FAB*	$129	1.0	$145	1.5	$278	1.5	$312	1.4
SSMC	$0	0.0	$82	0.8	$251	1.4	$293	1.4
HeJian	$0	0.0	$0	0.0	$176	1.0	$292	1.4
Grace	$0	0.0	$0	0.0	$140	0.8	$224	1.0
Jazz	$0	0.0	$141	1.4	$220	1.2	$212	1.0
Tower	$105	0.8	$52	0.5	$126	0.7	$186	0.9
ASMC	$62	0.5	$89	0.9	$139	0.8	$170	0.8
Silterra	$0	0.0	$68	0.7	$150	0.8	$165	0.8
Episil	$213	1.7	$189	1.9	$149	0.8	$139	0.6
CSMC	$34	0.3	$41	0.4	$80	0.4	$115	0.5
Winbond	$358	2.8	$74	0.7	$30	0.2	$20	0.1
Others	$134	1.1	$284	2.8	$385	2.1	$340	1.6
TOTAL	**$12,581**	**100.0**	**$9,994**	**100.0**	**$17,981**	**100.0**	**$21,518**	**100.0**

Note: * Includes 1st Silicon from 2000–2006 (est).

Source: IBS

Figure 2.2 Foundry Service Provider Revenues & Market Share

	TSMC	UMC	CHARTERED	SMIC
Market Share	49%	19%	7%	7%
Revenue Per Wafer ($)	1,335	1,073	1,115	904

Source: Company Reports

Figure 2.3 2006 Foundry Revenue Per Wafer ($M)

2.2 Semiconductor Back-End Services

Back-end manufacturing services can range from packaging characterization, substrate design, wafer bumping, wafer sort to assembly and final test, and may also include flip chip bumping services as shown in below in Figure 2.4.

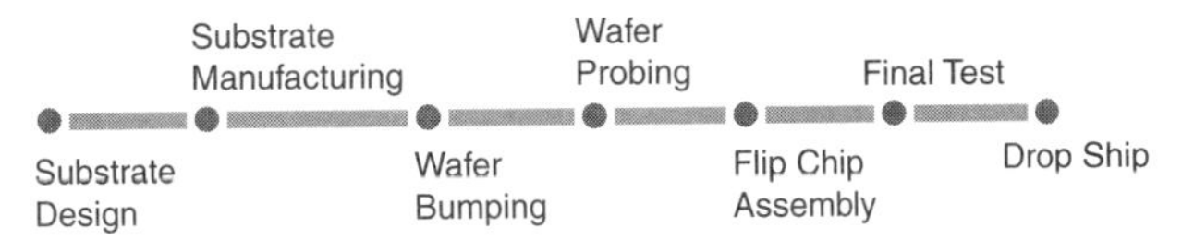

Figure 2.4

On the assembly and test side, the same three types of semiconductor companies exist: those with internal facilities, those that outsource and specialty assembly and test outsourcers. Very few integrated device manufacturers (IDMs) perform 100 percent of the packaging themselves, as the number of packaging types in the industry numbers in the multi hundreds and most IDMs are unwilling to invest in all the necessary equipment. Further, many IDMs (particularly high-performance IC manufacturers) focus on leading-edge packaging technologies in-house, but often leverage low-cost outsourcers in Asia for the mainstream processes. Still others (particularly low-cost producers) perform most of the bulk processes in-house but may contract out some of the higher-performance packages.

Test functions are often conducted in the same facility as packaging and assembly. However, many fabless companies may outsource their device packaging, but will perform testing in-house; this allows for faster design flaw identification.

The outsourced assembly and test market is less consolidated than the foundry market, with key vendors being ASE Test, Amkor Technology, Siliconware Precision (SPIL), and STATS ChipPAC. Almost all test activity is conducted in the Far East, as it is more labor-intensive than front-end processes and also carries less stringent clean room requirements. ASE provided photos (Figures 2.5–2.7) of flip-chip, wafer bumping and wafer back-end assembly and testing technologies.

Figure 2.5 Flip Chip Technology (Courtesy of ASE)

Figure 2.6 Wafer Bumping (Courtesy of ASE)

Figure 2.7 300-millimeter Wafer Back-end Assembly and Testing (Courtesy of ASE)

The trend toward outsourcing has also benefited back-end providers, though not as dramatically as for foundries. Figure 2.8 shows the growth of semiconductor assembly test services (SATS) revenues compared to the overall semiconductor industry.

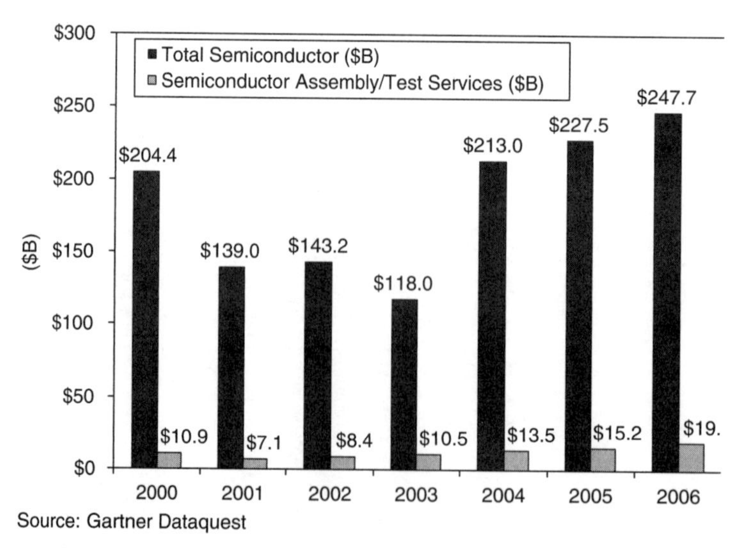

Figure 2.8 Back-End Services Growth Compared to Total Semiconductor Growth

The outsourced SATS market is less consolidated than the foundry market. The top two vendors in 2006 by revenue were ASE and Amkor, with a combined market share of 30 percent. Other key vendors include SPIL, STATS ChipPAC, and UTAC (Figure 2.9).

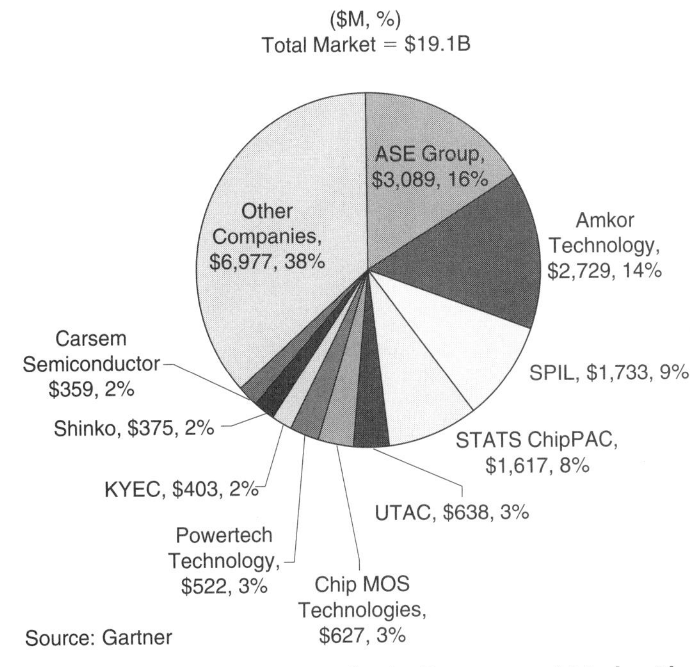

Figure 2.9 2006 SATS Companies by Revenue and Market Share

2.3 Semiconductor Equipment

The foundries and back-end manufacturers are dependant on their semiconductor equipment suppliers for setting up their manufacturing lines. Figure 2.10 is a list of the various semiconductor equipment suppliers categorized as front end (wafer manufacturing), back end (assembly, packaging and test) and process control and automation. For each category, the supplier's specialization has been provided.

Figure 2.11 provide the revenue estimates for 2006 for the major semiconductor equipment suppliers. Applied Materials leads the pack with a 2006 market share estimated at 15.2 percent, followed by Tokyo Electron with 10.5 percent, and ASML with 9.4 percent for a total net market of $42 billion in 2006.

In 2006, the semiconductor capital equipment industry totaled $40.5 billion, up from $32.9 billion in sales in 2005. Figure 2.12 shows equipment sales growth estimates for the industry to 2009, displayed by wafer processing, assembly and packaging, test and other categories of equipment. The global wafer processing equipment market segment increased 26 percent; the

FRONT END (WAFER MANUFACTURING)	Applied Materials	ASML	Tokyo Electron	Nikon	Canon	Novellus	Lam Research
Deposition	X					X	
RTP (Rapid Thermal Processing)	X						
CMP (Chemical Mechanical Planarization)	X						
Etch	X		X				X
Thermal Diffusion			X				
Photoresist Processing			X				
Lithography Scanners		X		X	X		
BACK END							
Assembly and Packaging	ASM	Seimitsu	Tokyo Electron	ASM	K&S	Unaxis	
Test	Advantest	Teradyne	Agilent	LTX	Yokogawa	Credence	
PROCESS CONTROL AND AUTOMATION	**KLA-Tencor**	**Brooks**	**Applied Materials**	**Asyst**	**HP**	**Daifuku**	**Hitachi**
Wafer and Reticle Inspection/Metrology	X						X
Defect Review			X				
Fab Automation		X		X	X	X	

Figure 2.10 Semiconductor Equipment Suppliers (Partial List)

2006 Rank	2005 Rank	Company	2006 Revenue Share ($M)	2006 Market (%)	2005 Revenue ($M)	2005-2006 Growth (%)
1	1	Applied Materials	$6,493.1	15.2	$4,738.5	37
2	2	Tokyo Electron	$4,481.7	10.5	$3,851.7	16.4
3	3	ASML	$4,004.1	9.4	$2,732.6	46.5
4	5	KLA-Tencor	$2,056.3	4.8	$1,654.9	24.3
5	7	Lam Research	$1,881.8	4.4	$1,147.0	64.1
6	4	Advantest	$1,794.0	4.2	$2,089.3	−14.1
7	6	Nikon	$1,519.2	3.6	$1,507.8	0.8
8	8	Novellus System	$1,389.1	3.3	$1,130.1	22.9
9	9	Dainippon Screen	$987.7	2.3	$991.3	−0.4
10	11	Canon	$924.3	2.2	$836.8	10.5
		Others	$17,106.3	40.1	$13,999.2	22.2
		All Companies	**$42,637.6**	**100**	**$34,979.2**	**22.9**
		OEM Elimination	$687.4	-	$538.0	27.8
		Net Market Total	**$41,950.2**	-	**$34,141.2**	**22.9**

Data includes revenue from acquisitions that occurred in 2006. 2005 data is before 2006 acquistions.

Source: Gartner Dataquest (April 2007).

Figure 2.11 Vendor Revenue Estimates for 2006

assembly and packaging segment grew 14 percent, the total test equipment sales increased 21 percent.

Figure 2.13 shows the semiconductor capital equipment market by region. While Japan continued to be the biggest spender for the third consecutive year at $9.2 billion in

Equipment Type	2005 $B	2006 $B	% Growth	2007 $B	% Growth	2008 $B	% Growth	2009 $B	% Growth
Wafer Processing	$22.9	$28.8	26%	$30.0	4.0%	$34.1	13.6%	$35.8	5%
Assembly & Packaging	$2.1	$2.4	14%	$2.5	4.4%	$2.9	14.7%	$3.0	5%
Test	$5.3	$6.4	21%	$6.6	2.1%	$7.5	13.5%	$8.0	7%
Other	$2.6	$2.9	10%	$3.0	3.4%	$3.3	9.1%	$3.5	7%
Total Equipment	$32.9	$40.5	23%	$42.1	3.7%	$47.8	13.3%	$50.3	5%

Source: SEMI

Figure 2.12 Equipment Sales Growth

2006, China's spending represented the highest growth rate at 74 percent ($2.3 billion) from 2005 to 2006, followed by the Rest of World combined (29.6 percent), North America (28.4 percent) and Taiwan (27.7 percent) all running neck in neck, but far behind China.

Region	2003 ($M)	2004 ($M)	2005 ($M)	2006 ($M)	2004 % Change from 2003	2005 % Change from 2004	2006 % Change from 2005
China	1,156	2,683	1,327	2,315	132.10%	−50.5	74.4
Europe	2,558	3,444	3,262	3,595	34.60%	−5.3	10.2
Japan	5,555	8,276	8,183	9,209	49.00%	−1.1	12.5
Korea	3,178	4,614	5,826	7,014	45.20%	26.3	20.4
North America	4,728	5,812	5,702	7,324	22.90%	−1.9	28.4
Taiwan	2,917	7,762	5,722	7,308	166.10%	−26.3	27.7
Rest of World	2,097	4,490	2,862	3,709	114.10%	−36.3	29.6
Total Regions	$22,189	$37,081	32,884	40,474	67.10%	−11.3	23.1
Source: SEMI-SEAJ							

Figure 2.13 Semiconductor Capital Equipment Market by World Region

CHAPTER 3

Qualities of Successful Fabless Companies

This section discusses the fabless semiconductor industry environment in terms of what the reader can expect from individuals and the fabless community as a whole.

3.1 Defining Events for the Fabless Market

The rise of fabless semiconductor companies and corresponding foundries started in the mid-1980s. At the same time, the industry shifted away from vertically integrated manufacturing toward a focus on core competencies. Companies began subcontracting activities they deemed lower in value, including a new emphasis on wafer manufacturing. The prevailing attitude was "silicon is free." Today, state-of-the-art semiconductor manufacturing is pushing up against the limits of physical and chemical process understanding.

The catalyst for this situation started in the early 1980s. Junk-bond financed-leveraged buyouts and corporate raider takeovers resulted in corporations that were torn apart because the sum of the constituent parts was worth more than the whole. The spillover effect was that corporations sought to increase their stock price through divestiture while shifting investments to areas that offered a competitive advantage. This trend challenged the long-held belief in market power gained through vertical integration. Metrics to benchmark these activities, such as revenue per employee, return on invested capital (ROIC) and economic value add (EVA) became fashionable in investors' eyes and contributed toward a trend of outsourcing capital-intensive activities.

Given these events, the uncoupling of semiconductor design and manufacturing became not only viable but also preferred for most semiconductor product areas. Some general exceptions were commodity like dynamic random access memory (DRAM), high-volume PC/embedded processors/DSPs and mixed-signal products.

3.2 Thriving in the Fabless Model

Successful fabless companies flawlessly execute a succession of enthusiastically adopted high-margin products. Their execution methodology effectively uses the optimum technology that wafer foundries can offer. However, before examining in more detail the common threads that run through successful fabless companies, it is important to understand the main categories of companies that have thrived in this model.

3.2.1 Application-Specific Standard Products

The rise of the Internet in the mid-1990s enabled one of the most explosive growth markets for networking equipment. This expansion, in turn, drove the corresponding growth of networking application-focused fabless semiconductor companies as well as the acquisition of many fabless start-up companies. Specific areas include Ethernet physical layer and switching products, broadband access devices such as digital subscriber line (DSL) and cable modem chipsets and telecom data path devices such as SONET framers and network processors. Companies such as AMCC, Broadcom, Centillium, GlobespanVirata (acquired by Conexant), Marvell, and PMC-Sierra combined their unique algorithms and design capabilities to deliver silicon that supported this revolution.

3.2.2 Programmable Logic

The leading programmable logic companies, Actel, Altera and Xilinx, started with a simple idea: most designs were not high-enough volume to justify the non-recurring engineering (NRE) costs and risks of a custom chip. Today, this market has blossomed into one of the most successful examples of the fabless model at work and has thrived because of competition and the fact that no one company has completely dominated the market.

3.2.3 High-Performance Processors

The fabless model has dramatically facilitated entry for delivering leading-edge microprocessors. SiByte (acquired by Broadcom) and QED (acquired by PMC-Sierra) delivered MIPS processors with best-in-class performance. Nvidia and ATI Technologies (acquired by AMD in 2006) continue to battle for dominance in the graphics processor market, each utilizing the most advanced fab process technology on par with the ×86 processor it sits next to. The fabless model for processor companies is continually being validated as new-generation, specialized processors from Intrinsity and Xelerated are pushing the envelope on both circuit design and processor architecture.

3.2.4 High-Speed SRAM

Today's high-speed SRAM and content addressable memory (CAM) products are not a commodity like DRAM. The DRAM market is rapidly concentrating to a few dominant suppliers who own their fabs. SRAMs and CAMs are niches that have experienced an explosion of interface types, configurations and packaging options and are pushing state-of-the art for logic process technologies from foundries. Companies such as Alliance Semiconductor and Integrated Silicon Solution, Inc. (ISSI) have thrived in this environment and have added a twist – foundry partnerships. Both companies have made significant financial investments in their respective foundries as part of overall programs for cooperative R&D and wafer supply guarantees.

3.3 Key Qualities for Success

Still it must be more complicated than simply delivering the best designs in each of these respective categories. Successful fabless companies all possess a set of common key qualities.

3.3.1 Essential Market and Customer Understanding

The biggest challenge for start-ups and large companies alike is truly understanding customer needs and market problems and then defining the right products to fill those needs. As the cost of mask sets has risen in excess of $2.5 million for 65-nanometer technology (Figure 3.1), understanding needs and responding correctly becomes even more critical as the probable return for a given investment must be even greater than in prior product generations. While it seems obvious that any company would do the research before it spends upward of $25 million to $40 million or more to develop a new product, successful companies clearly target large markets where they can be the number-one or -two player and earn required returns – a positive EVA.

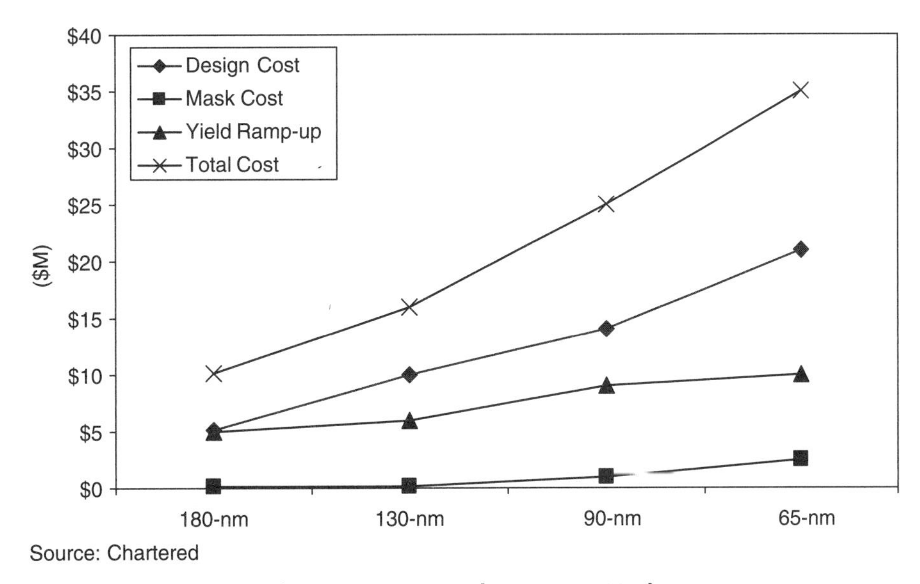

Figure 3.1 IC Cost by Process Node

Many systems markets that consume semiconductors have consolidated and have become more concentrated with a few large players. This effect became more pronounced in the networking industry, for example, after the downturn in 2001, which was precipitated by too much inventory and the dot.com bust. Weaker players and start-ups, in particular, have experienced a difficult time. For example, in the edge router market that Cisco continues to dominate, Juniper acquired the number-three player, Unisphere, to strengthen its number-two position, and more recent entrants have been acquired or struggled and eventually ceased operations.

Thus, understanding market requirements often equals understanding the system needs of major players in their respective markets. This concept must be translated into revenue by engaging and winning market leaders early in the design cycle. These design wins need to be in markets that have enough growth and longevity to generate sustained revenue and returns. In addition, as an increasing amount of system design is being pushed onto semiconductor makers, being the primary source on a reference design can increase revenue potential greatly for platforms that have become widely adopted by market-leading original equipment manufacturers.

3.3.2 Relentless Focus on Execution

While it is necessary to have essential market understanding, fabless companies cannot succeed without execution. Tools and methodologies are keys to enabling a team of well-managed, talented engineers to execute successfully.

Today's design tools enable rapid verification and timing (RVT) closure. Logically verified RVT can be taken to tapeout in a matter of weeks. The continued advance of electronic design automation (EDA) tools and libraries enables fabless companies to execute processors, logic devices and advanced memories with up to a hundreds of million of transistors in today's advanced process geometry. In addition, foundries offer limited "shuttle" services that allow multiple die from different vendors to be shared on the same reticle for a fraction of the cost of a dedicated mask set. This feature can enable key pieces of the design to be evaluated and debugged before committing to a dedicated mask. Finally, today's complex designs can often be verified in hardware using field programmable logic devices (FPLDs), which themselves are delivered using the fabless model.

3.3.3 Relentless Focus on Costs

A Silicon Valley adage states that one can predict the downfall of a rising company as soon as it builds a palatial campus. While this correlation tends to be exaggerated, it does reflect that these companies were unable to achieve the next level of growth when they lost focus on driving out costs. Given that the fabless model enables a huge fixed-cost burden to be shifted to the foundry partner, fabless semiconductor companies naturally seem well suited to manage costs.

While a focus on driving down costs is influenced by many factors, it is important for fabless companies to work closely with their foundry partner to use a process technology that optimally balances die cost, performance and technology risk when the design ramps to production. For new designs, missing a market window means the difference between success and absolute failure, so process technology risks must be minimal, even if the yielded die cost may not be the lowest. Successful market penetration can be the springboard for die shrinks along with feature and performance enhancements. On the other hand, for more commodity-driven markets, planning a roadmap for cost reduction is a requirement that aggressive cost reductions are planned based on a clear understanding of the foundry's technology roadmap.

3.3.4 Management

An endless stream of management books continually dispense common-sense wisdom on the subject of management. However, with the fabless semiconductor company, two primary functional areas must work together for success, regardless of the strength of the chief executive officer (CEO), the other key executives or the manufacturing partners: engineering and marketing. The synchronization of these two disciplines results in execution of the right products at the right time for the market.

Marketing builds and fosters customer relationships to gain insight into true market requirements that it uses to develop a sustainable chain of products that propel the company. Technical-level relationships, on the other hand, address the "devil's in the details" to ensure exhaustive coverage of customer requirements and the right architectural tradeoffs that guarantee product execution and time-to-market. Failure to synchronize this process between marketing and engineering results in products that miss the market window and market requirements. Synchronization is vitally important in a ruthlessly competitive environment, because being almost right is often no better than being completely wrong.

3.4 The Future of Fabless

Without getting caught in the quagmire that has plagued prognosticators over years past, it should be recognized that innovation lives on, and the future of the semiconductor industry is closely tied to the future of fabless companies. The fabless model will continue to thrive as long as fabless companies leverage the model to focus on the key qualities for success. The model gives them the best opportunity to deliver the right products at the right time and meet the challenge of thriving in a fluid market that presents an ever-increasing amount of uncertainty and promise.

PART 2

An In-Depth Understanding of the Fabless Semiconductor Business Model

CHAPTER 4

Semiconductor Manufacturing Basics

The wafer manufacturing process is the reason the fabless business model was created years ago and is gaining strength each year.

Semiconductors are manufactured in specialized factories, or wafer fabrication facilities. Fabless companies outsource this process to wafer foundries, which own fabs equipped to manufacture a fabless company's chips that are developed on the wafers. IDMs may have their own fabs to manufacture these wafers, or they may choose to outsource this process to foundries, as the industry is seeing this trend increasing each year.

4.1 Semiconductor Processes

4.1.1 CMOS

Complimentary metal oxide semiconductor, most commonly referred to as its acronym CMOS (pronounced "see-mos"), is the most used semiconductor design process type in production today. The CMOS process uses standard silicon wafers and combines PMOS (positive) and NMOS (negative) transistors in certain ways, so that the final product consumes less power than PMOS-only or NMOS-only circuits. In general, devices fabricated using a CMOS process will be less expensive to manufacture, as well as consume less power than devices fabricated using other processes.

4.1.2 Specialty/Compound

There are a wide array of specialty processes and devices. When connected to a power source, compound semiconductors can perform highly specialized functions – especially those involving certain compounds more suited to specific types of applications, such as high-speed communications or power management and amplification.

Silicon Germanium (SiGe) is a commonly used compound for high-speed physical layer devices in wireline communications (e.g. 3G and wireless, digital set-top boxes, automobile radar systems, personal digital assistants (PDAs)).

Gallium Arsenide (GaAs) is a direct bandgap compound semiconductor material with inherent high electron mobility property (6x of Silicon). GaAs is often used in wireless handsets and set-top boxes, as it is better suited to radio frequency (RF) applications. It is also used in optoelectronic devices for wired communications.

Indium Phosphide (InP) is used in high-power and high-frequency electronics because of its superior electron velocity with respect to the more common semiconductors silicon and GaAs. It also has a direct bandgap, making it useful for optoelectronics devices like laser diodes (e.g. Optical networking, digital instrumentation, microwave communications).

BiCMOS is the integration of bipolar junction transistors and CMOS technology into a single device. This technology has commercial application in amplifier and discrete component logic design (e.g. Wireless applications).

Gallium Nitride (GaN) is used in light-emitting diodes (LEDs) and laser diodes (LDs). Ideal for high working-temperature and high-power, high-frequency RF devices (e.g. RF power transistors).

Several advantages of compound semiconductors include superior performance characteristics over silicon. For example, including higher switching speeds allow for greater bandwidth; lower power consumption results in longer battery life; and less heat generation increases a unit's functionality.

4.2 Semiconductor Manufacturing Steps

1. *Wafer Production*: Prior to wafer fabrication, polycrystalline silicon that contain dopants that can modify conductivity is melted down. From the melted silicon, ingots are grown and then shaped and sliced into thin wafers so the process can begin.
2. *Wafer Processing*:
   a. *Thermal Oxidation*: The wafers are pre-cleaned using high-purity de-ionized water and various low-particulate chemicals, a must for high-yield production. The silicon wafers are heated to approximately 1,000 degrees C and exposed to ultra-pure oxygen in the oxidation furnace.

   b. *Photolithography*: This process is used to transfer circuit patterns onto the wafer. Beams of light are projected through a patterned reticle (mask) onto the wafer that is covered with a photosensitive material that etches a circuit into the wafer to expose the resist. This process is often compared to photography and creating images from film.

   c. *Etching*: Materials are removed during several steps using multiple tools. The beam of light from the photolithography "hardened" portions of the wafer as a result of the resist (mask) that was applied in the earlier step. This "hardened" portion is no longer sensitive to the light. The "non-hardened" resist washes away during the development process, and the material below it is etched away. It is then exposed to a chemical wet solution or plasma (gas discharge) so that areas not covered by the hardened photo resist are etched away. The "hardened" resist is then stripped away using either wet or plasma chemistry, forming a three-dimensional pattern on the wafer. The first photolithography etch process will result in a pattern. Before moving to the next step, the wafer is optically inspected to assure that the image transfer from the mask to the top silicon layer is correct. There are often several lithography and etch steps when manufacturing a wafer. Subsequent layers of various patterned materials are formed on the wafer to create the multiple layers of circuit patterns on a single chip.

d. *Doping*: The doping process controls the flow of electricity through the chip. It allows for certain areas of the wafer to be exposed to chemicals that change its ability to conduct electricity. Atoms with one less electron than silicon (such as boron) or one more electron than silicon (such as phosphorus) are introduced into the area exposed by the etch process, to alter the silicon's conductivity. These areas are called "P type" or "N type", respectively, which reflects their conducting characteristics. Interconnections are formed at the portions of the chip where electricity is conducted. A metal (usually copper) is then electroplated on the entire wafer surface. Metal can be chemically and mechanically polished away if it is unneeded. All of these metal interconnects form pathways that must connect for the chip to function properly.

3. *Dicing*: First, the wafer is run through a wafer sort process, which detects the electrical performance of each chip on the wafer. The chips that fail are marked and discarded later after a diamond saw or laser separates them. The remaining chips are visually inspected under a high-power microscope before being packaged.

4. *Assembly/Packaging*: For protection and functionality, each die is first attached to a package, and wire bonding is used to connect the input/output gates on the die to the leads on the outside of the package. The device is then encapsulated, sealed and markings are affixed. A wire bonding machine attaches wires, a fraction of the width of a human hair, to the leads of the package.

5. *Test*: Once the chips are packaged, they each go through another round of testing prior to delivery to the customer.

4.3 Wafer Size

Wafers come in different sizes, which are determined by their diameter. Over the years, wafer sizes have increased to allow more die to fit on the surface, resulting in reduced manufacturing cost. The cost factor is the sole reason companies choose to use larger wafers. No other advantages exist today, as a larger wafer does not improve chip performance for the semiconductor company or provide any other benefits.

The reason the cost savings is so great is because a semiconductor company can increase the number of die on a single wafer by 2.25x simply by using a wafer 50 percent larger in diameter.

Moving to larger wafer sizes includes equipping fabs with new equipment, which can be costly to a fabrication facility. Transitioning to larger wafers only occurs every few years, and the time gaps in between sizes are increasing with each move. Many of today's leading foundries have the capabilities to manufacture 200-millimeter (8-inch) and even 300-millimeter (12-inch) wafers. The move to 450-millimeter (18-inch) wafers will not likely occur until the 2010–2015 timeframe.

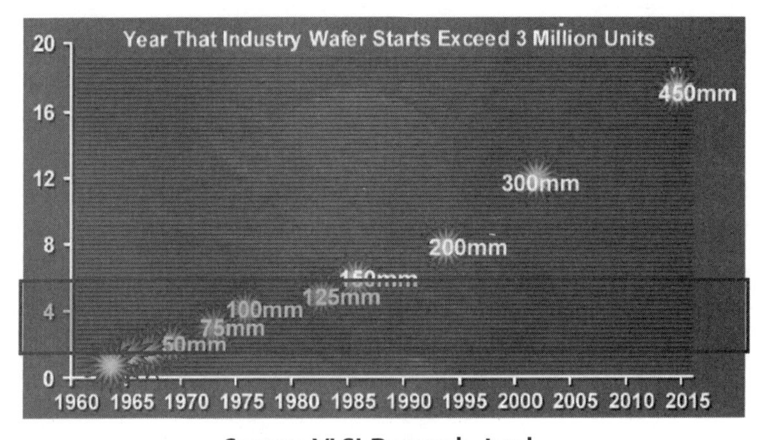

Source: VLSI Research, Intel.

4.4 Manufacturing Costs

Both direct and indirect costs must be considered when owning and maintaining a fab. Direct, fixed costs include:

- Real estate
- Facilities
- Taxes
- Plant depreciation
- Equipment depreciation

Variable costs include:

- Raw-wafer costs
- Consumables

- Labor: operators, process engineering support, equipment engineering and maintenance
- Equipment: production hours/year, maintenance costs and operating costs
- Utilities

These costs can add up for semiconductor companies that own fab facilities. Based on these, plus other factors, many companies are now allowing the pure-play foundries to incur the costs.

4.5 Conclusion

As process nodes shrink, the wafer manufacturing process experiences greater challenges to overcome, including climbing costs. As a result, more companies currently owning fabrication facilities are finding that it makes sense to outsource the fabrication process to pure-play foundries instead of equipping and maintaining older fabs with expensive, state-of-the-art equipment.

CHAPTER 5

Fabless ASICs

5.1 Introduction

The semiconductor industry has a long history of business model evolution and innovation, the fabless semiconductor model itself being perhaps one of the most significant transformations. And, even within the fabless model, there have been a variety of permutations and adjustments, usually dictated by some macro-economic trend that forces companies to think differently about how they can most efficiently get products to market. It is certainly well accepted now that owning and operating one's own fabrication facility is not only unnecessary, it is, in most cases, untenable financially. That same focus on efficiencies and economies of scale has permeated many other aspects of the semiconductor value chain, resulting in a further refinement of the fabless semiconductor business model: the fabless application-specific integrated circuit (ASIC) model.

5.2 Origins of the ASIC Industry

The ASIC industry emerged in the early 1980s as a way to provide companies, specifically system original equipment manufacturers (OEMs), access to the design tools and semiconductor technologies necessary to implement application-specific devices for their systems – without the need for the significant investments associated with the semiconductor business. This enabled these companies to focus on what they knew best – the needs of their customers and their specific markets – and let the ASIC suppliers be the experts in the design and manufacture of semiconductors.

Because the required technology was not readily available from third parties, early ASIC suppliers were forced to develop all of the technology themselves: design tools, libraries, IP cores, process technology – even the fab itself. However, during the 1990s, the emergence of the fabless semiconductor industry, including a full ecosystem of suppliers, enabled all of the required elements to design and manufacture a chip to be sourced from third parties: electronic design automation (EDA) tools, cell libraries and IP blocks on the design side; and package, assembly and test services and wafer fabrication on the manufacturing side.

5.3 Emergence of the Fabless ASIC Business Model

The availability of the required technology enabled the emergence of a new model for the ASIC business: the fabless ASIC model. The essence of this model is to provide the same solution to a company as the "classic" ASIC suppliers have traditionally done – but to leverage the fabless semiconductor ecosystem to obtain the required technology. Rather than developing everything in-house, the fabless ASIC model utilizes an outsourced approach, which has inherent advantages over the traditional ASIC model.

Fabless ASIC companies minimize their financial investment by relying on third parties to develop the underlying semiconductor technology, as well as lower their total production pricing through lower operating margins, due to their reduced need for significant R&D investments. They also have access to the full range of capabilities available from the entire semiconductor supply chain versus relying on the limited range of capabilities they have developed themselves, providing the flexibility required to truly support the needs of their customers.

- Due to its inherently lower cost structure, the fabless ASIC model has also opened the door to a new class of ASIC customers: fabless semiconductor companies. While system OEMs compare the fabless ASIC model to the "classic" ASIC model when searching for a solution, fabless semiconductor companies typically measure the fabless ASIC model against a direct engagement to the semiconductor supply chain. For fabless semiconductor companies, a fabless ASIC supplier enables them to:
  - Control development costs by only using resources on an "as needed" basis
  - Eliminate the cost, complexity, ramp-up and low return rate of implementing a customer-owned tooling (COT) approach for back-end design
  - Obtain access to the most skilled resources for a particular job, without the complexity and financial overhead of finding and hiring skilled engineering personnel
  - Achieve an optimized cost structure by leveraging the buying power of the aggregated demand of the fabless ASIC supplier's multiple customers
  - Attain more predictable pricing, both short term for non-recurring engineering (NRE) expenses and long term for production costs
  - Lower risks by utilizing a proven, predictable production partner – with the highest possible yields and lowest overall costs.

5.4 The Fabless ASIC Model: How It Works

From a revenue and profitability point of view, the fabless ASIC model can be broken down into two elements. The first element consists of the charges incurred during the design of a device, the NRE, which include the costs to do the design, to procure the third-party IP, and to obtain the mask set, prototype wafers and test hardware. A fabless ASIC company typically does not seek to make a profit on these activities, and in fact often discounts this expense to its customer – the fabless semiconductor company or system OEM – to secure the rights to build the device in production-volume quantities.

The second element is the revenue associated with the production of the device, which is where a fabless ASIC company makes its profit. A fabless ASIC company is able to offer a competitive price for the production of its customer's semiconductor device, while at the same time producing the device for a cost which is lower than the price. It is able to do this for a number of reasons, the most significant of which are the efficiency of the fabless ASIC company's design and productization efforts, which result in the lowest possible device cost; and the low costs of the materials, which are achieved through the fabless ASIC company's aggregated supply chain purchases of its many customers. The purchasing power enabled by this supply chain aggregation enables significantly lower material costs than its customer would be able to obtain on its own.

5.5 The Services and Capabilities of a Fabless ASIC Supplier

A fabless ASIC company manages the entire chip development and manufacturing process and applies resources when and where its customer needs them. For the typical fabless semiconductor company, which may only put one or two designs through to production a year, access to this type of capability can mean the difference between success and failure – especially if the only alternative is to hire, train, maintain and manage the required engineering resources internally.

To successfully implement a chip design with the size, degree of functional complexity, mix of design technologies and manufacturing challenges afforded by leading-edge processes requires the integration of an unprecedented number of technological and managerial specialties. Despite the technical advances offered by each supplier in the semiconductor supply chain, managing a large number of suppliers who ultimately contribute to the design and production of a complex device is an increasingly challenging task. As a result, the fabless ASIC model has emerged as the best approach to dealing with the escalating costs and complexity of designing and manufacturing a state-of-the-art chip. The following sections will describe in more detail the services, capabilities and benefits of working with a fabless ASIC company.

5.5.1 Physical Design

All contemporary chip designs targeted at the 90-nanometer process node or beyond require a sophisticated set of capabilities to successfully implement them in silicon. Most designs implement a number of functions and contain millions of gates of logic, megabits of memory and complex elements such as analog circuits or multi-gigabit per second I/O interfaces. Their complexity earns these device designs the description of systems-on-chips (SOCs).

The industry has traditionally divided the design cycle into two phases: front-end design, comprising the architectural, system register-transfer-level (RTL) design and verification and logical synthesis phases; and back-end design, often called physical design, including floorplanning, placement, routing, timing closure and physical verification.

The responsibility for the front-end design typically falls on the fabless semiconductor company or system OEM, as this phase of the SOC development requires a deep understanding of the end-application that the device is going to serve. In fact, the architecture and RTL implementation of the device is usually the core differentiating aspect of the company's design.

Conversely, back-end design is often left to an external supplier, such as a fabless ASIC supplier, for a number of reasons: smaller companies lack the scale to justify the required investment in tools and engineers, and even larger companies recognize that physical design is neither their area of expertise nor an activity that adds value to the solution they provide to their end-customers.

Fabless ASIC suppliers are well suited to assist these companies with the physical design of their chips for a number of reasons. First and foremost, this is their business and they do it day in and day out. They possess the tools, knowledge and capabilities to successfully transform a company's SOC design into a working piece of silicon, and the high volume of designs they execute enables them to become experts in the required tools and methodologies. They also have experts on staff to deal with the full range of technical considerations that may be encountered in a design. By amortizing their tools, staff and infrastructure across a large number of their customers, they are more efficient operationally than any individual company attempting to develop an SOC device on its own.

Second, the interests of fabless ASIC companies are inherently aligned with their customers' interests. Specifically, the fabless ASIC model requires volume production to produce a profit for fabless ASIC suppliers. As a result, their main motivation is to enable companies to get a working chip in the shortest time possible. This is in stark contrast to a design services company, whose only interest is to get paid for the design effort. In fact, the longer it takes to complete the design, the more money the design services provider can make.

Another consideration is the increasingly close interaction between the design techniques used in implementing an SOC design and the manufacturability of the device. Issues such as design rules and circuit performance have always been considered when doing the physical design of a chip, but in more advanced process nodes the considerations are much more subtle – cross-talk effects and antenna effects, for example. Because of their expertise in this area and the fact that they participate in the production of the device, fabless ASIC companies are well positioned to act as intelligent intermediaries between design and manufacturing.

5.5.2 Selecting and Integrating IP into an SOC Design

To decrease development time and increase design quality, almost all SOC designs make use of reusable IP blocks for common functions ranging from microprocessors to peripheral controllers, from DSPs to ADCs and DACs.

The use of these IP blocks has become so common it has lead to the creation of a new industry: third-party IP suppliers that specialize in developing IP cores for specific functions. Unfortunately, it is often the case that much of the increased productivity that is expected through the use of off-the-shelf, third-party IP is lost in the management of the IP supplier and the integration of the IP into the design.

A fabless ASIC company, through its expertise in the relevant technical areas, its relationships with third-party IP providers and its experience with using these IP blocks in multiple customer designs, can greatly improve its customers' experiences in this area. It can assist the customer in selecting the most appropriate, lowest risk and most cost effective IP cores, and work with the customer to successfully integrate these blocks into the customer's design. Should a problem arise, the fabless ASIC supplier can again call on its internal technical expertise and supplier relationships to determine the cause of the problem and implement a solution.

5.5.3 Package Design

Virtually all semiconductor devices are incorporated into one form of package or another. These range from simple lead-frame packages for low-speed, low-signal count devices, to complex multi-layer flip-chip packages for high-power, high-signal count and high-frequency devices. In addition, the interrelationship between the die and the package is becoming increasingly complex as the limits of signal performance and power are pushed.

To address this situation, the fabless ASIC company can also provide package design services. This allows its customers to benefit from the synergy of having one supplier develop both the die and the package, and, as a result, take ownership for the interactions between the two elements.

5.5.4 Design for Testability and Production Test

Another important consideration during the design of a device is how it will be tested in production. It is critical that the designer of an SOC device take test considerations into account during the design process to ensure adequate production quality and cost. On the one hand, inadequate test coverage can be disastrous for a device shipping in high volume.

A test hole which causes a supplier to ship parts with a few hundred defects per million units can result in significant financial exposure and damage to their reputation. On the other hand, over-testing or testing inefficiently can result in a product cost structure which is unacceptable. Fabless ASIC suppliers can assist companies in solving this through their expertise in design, design for testability (DFT) tools and methodologies and production test for high-complexity, high-volume devices.

Another service provided by a fabless ASIC company is the development of the production test program for the device. This effort involves defining a test plan for the product; selecting the target test platform based on the specific requirements of the device to be tested, such as the number of pins; speed at which test is to be performed and the analog content in the device. Again, because of its experience in servicing a broad range of customers, as well as its relationships with multiple test suppliers, a fabless ASIC supplier is a very capable partner in this effort.

5.5.5 Characterization

In spite of the new analysis and verification tools provided by EDA tool vendors, designers still need to verify the actual silicon. Ultra deep sub-micron manufacturing cannot avoid process variations, not only from lot to lot, but also from wafer to wafer, die to die and even transistor to transistor on the die.

To address this, actual silicon needs to be characterized over the full range of process variations expected to be seen in volume production. Corner or "split" lots of material are processed by the semiconductor fab and tested over the range of temperature and voltages under which the part is required to operate. The characterization corner lot splits are designed to stress the worst-case process extremes that could be observed in a production ramp. In this process, any sensitivities of the design to certain process corners can be identified and addressed before taking the design into volume production. Methods of addressing such sensitivities include shifting some process parameters to avoid the particular point in the process that the design is sensitive to, changing the test program to relax a design parameter that is defined too stringently or in severe cases, modifying the design.

5.5.6 Quality and Reliability

In addition to characterization, qualification of designs is also necessary. Qualification testing ensures that the SOC device will not be damaged during the board-level manufacturing process, and that it will continue to perform as specified during the required lifetime of the product into which it is incorporated. Tests done during qualification include electrostatic damage or discharge testing, where the device pins are subjected to high voltages to ensure they will not be damaged during manufacturing; latch-up testing, to test the susceptibility of a short between power and ground rails triggered by input or output current or supply over voltages; and accelerated life testing, which by applying higher voltages and temperatures to a sample of devices, can predict the expected lifetime of the device in normal operation.

The level of qualification testing required is usually determined by the end-customers and applications that the device targets. Low-cost, short lifetime, consumer devices may be able to accept a fairly minimal qualification effort, but a large telecommunications system device, which requires high reliability in harsh environments for 10 years or more, will demand a much more thorough and extensive qualification process. Since fabless ASIC suppliers serve a wide range of companies and markets, they are ideally suited to guide the designer of a complex SOC device through the qualification process to meet the demands of the end-customer.

In addition to the qualification of the specific design, a fabless ASIC supplier can also support the quality infrastructure that fabless semiconductor companies and system OEMs require. This includes supplier qualification, review and monitoring; document control and process change notifications (PCNs).

5.5.7 Volume Manufacturing

Once a device is released for manufacturing, it is important to achieve a quick and cost-efficient ramp up to volume production. If done properly, characterization and qualification, as described, can significantly smooth the production ramp for a product. However, despite this proactive effort prior to production ramp, ongoing support is still required during production. The benefits associated with using a fabless ASIC company during volume production include:

- Pre-defined unit price
- Lower material costs through the aggregation of purchases over a broad base of companies
- Supplier management
- Yield management
- Ongoing cost reductions
- Material planning and order placement
- Management of the material flow and logistics
- Die banking and just-in-time (JIT) delivery programs to increase delivery flexibility
- Warrantee on finished units
- Financing the WIP

- Processing of field returns (returned materials authorizations, RMAs) and failure analysis to determine the root cause of the failure
- Information technology (IT) infrastructure such as enterprise resource planning systems and work-in-process tracking systems

The net result is that companies utilizing a fabless ASIC supplier to manage their volume production will enjoy lower costs and lower risks than they would experience if they engaged directly with the semiconductor supply chain.

5.6 Conclusion

The evolution of the semiconductor industry has as much to do with the emergence of new business models as it does with technological innovations. Given the complexities and expenses associated with successfully and cost-effectively producing a leading-edge chip, there is a constant need to evaluate new approaches. The fabless ASIC model is ideal for anyone seeking to develop a custom IC: fabless ASIC companies are able to coordinate the efforts of the multitude of suppliers who contribute to the development of a complex chip; to provide technical expertise in chip design and manufacturing and to enable purchasing leverage through the aggregation of the semiconductor supply chain. These capabilities are combined with a business model which perfectly aligns the motivations of fabless ASIC companies with their customers' urgency to get working silicon to market as quickly as possible, with the lowest cost.

In the increasingly complex world of custom chip development, more and more companies, be they a system OEM or a fabless semiconductor company, will look to a trusted partner to oversee the entire supply chain in both the design and manufacturing phases of their SOC development, enabling them to focus on their core value-add of developing unique approaches to system design. That trusted partner is a fabless ASIC company.

CHAPTER 6

Electronic Design Automation

6.1 Fabless EDA Overview

Electronic design automation, or EDA, is a collection of software tools used for chip design, verification and validation. EDA helps to facilitate and automate the process of electronic design.

Customers of EDA products and services include major integrated circuit (IC) system design companies that create products for the computer, networking and consumer markets. This market can be logically extended to include any company involved in electronic design.

EDA helps customers solve design challenges by providing leading-edge electronic solutions that streamline the process of moving advanced IC and system designs to volume production. Customers use EDA software and hardware methodologies and services to design and verify advanced semiconductors, printed circuit boards (PCBs) and systems used in consumer electronics, telecommunications equipment, networking devices and computer systems.

Figure 6.1 lists the leading EDA companies as of 2006:

	2005 Sales ($M)	2006 Sales ($M)
Cadence Design Systems	$1,329.2	$1,483.8
Synopsys	$991.9	$1,095.6
Mentor Graphics	$705.2	$791.0
Magma Design Automation	$155.7	$172.1
Source: Company reports		

Figure 6.1 EDA Companies 2006 Sales

Emerging EDA companies include:

- Clear Shape Technologies, Inc.
- Apache Design Solutions
- Jasper Design

- Berekely Design
- Sierra Design Automation
- Calypto Design Systems
- Atrenta

6.2 Fabless EDA Selection Process

The selection process is based on both business and technical criteria, depending on the type of design. Determining factors can include whether the design is analog, digital, or mixed-signal; the complexity of the design; and whether they are designing the package, the board or the system. Many people who work at fabless companies have had experiences with different EDA solutions in the past, which is another contributing factor in the selection process. Companies involved with less-complex designs will have more limited needs, whereas clients involved in complex, leading-edge technologies may require extensive engineering consultations with their supplier.

The challenges faced by companies working with EDA suppliers include time-to-market, managing design complexity and integration for value. Today's consumers – who are customers of the EDA industry's clients – want to put a supercomputer in their pocket. This demand requires the design of products with multiple fabrics, including digital, analog, radio frequency, system-on-chips (SOCs), system-in-packages (SiPs) and printed circuit boards (PCBs), as well as the resolution of conflicting technical and business objectives that must be simultaneously addressed. From a technical standpoint, issues such as low-power design and signal integrity must be addressed. From a business perspective, design-for-manufacturability (DFM) issues such as yield and yield ramp (which drives profitability), are paramount. EDA tools are created to address this diverse range of issues and solve the associated challenges for the customer.

EDA is the inception point of electronic design, and EDA tools offer improved design team productivity. EDA tools help design teams meet the intense challenges presented by Moore's Law, enabling them to deliver faster, smaller, more powerful, and more power-efficient designs in a time-efficient manner. These design teams need to verify their concepts, model and analyze their designs and identify and eliminate problems before going to production. EDA helps get it done right.

Designers first need EDA tools to specify their design intent and then implement the design. Verification tools are used during various stages of this process to ensure the design will perform as intended. The design may need to be implemented using both digital and analog or mixed-signal tools, and during implementation, verification tools will be run multiple times.

After implementation, the development team must consider how the design will fit into the package, and how the package will fit into the board itself. Once the design is finished, the IC is handed off to the manufacturer. As a part of this process, the team works to resolve design for manufacturing issues that may occur, and to ensure the design can be quickly ramped up to volume production.

The fabless companies assess the type of design they are creating, including the complexity of the design and the targeted end market. Based on this assessment, fabless companies select the appropriate tools and services from the EDA suppliers.

Addressing today's advanced design challenges requires highly integrated solutions, including reference flows and design kits. Leading nanometer design teams working on complex designs rely on integrated solutions and close partnerships with their EDA tools supplier.

Most EDA tools are interoperable, and fabless companies must weigh the trade-offs between using an integrated solution and mixing and matching tools from different suppliers. In cases that use design tools from multiple sources, additional resources and investments may be required to create an integrated design flow. EDA has many well-defined and de facto standards, especially in the areas of timing models, databases, layout formats, and high-level modeling languages.

EDA is the bridge between manufacturing and design. It provides the vehicle for communicating process information from the manufacturer to the designer in a manner that is compatible with the EDA tools, and enables the designer to evaluate the impact of design manufacturability issues in advance of going to silicon. The primary data that fabless companies need are device models, interconnect models and design rule checks (DRC)/LVS technology files. In addition, companies engaged in analog or other forms of custom design will need process design kits (PDKs) from their manufacturer or EDA suppliers.

The following presents an overview of the various factors that need to be considered during the EDA selection process.

6.2.1 Device and Interconnect Models

On an IC, the two main components are devices or transistors, and interconnect or wires which connect these transistors. Device models are created for transistors and can be used for analysis during the design phase. Similarly, interconnect models represent the wires that connect the devices, and can be used to measure the delay in signal transmissions or the coupling and noise characteristics between different signals.

Device models can be integrated with interconnect models, delivering a method for predicting the behavior of the circuit before hardening it in silicon. This synergistic approach shortens development cycles and enhances the productivity of the design teams. The silicon manufacturers provide these device and interconnect models as part of their services.

New spatial effects have emerged in nanometer technologies that are having an impact on device models and interconnect. This impact results from both proximity effects and material effects. When copper was adopted, proximity effects changed the behavior of the interconnect; manufacturing techniques are impacting how the thickness of the interconnect (wires) varies across the die and across the wafer. These variations are based on the neighbor's reference neighborhood, including other interconnects that may be nearby.

On the device side, proximity effects have created the need for new kinds of device model requirements. These requirements are driving improvements in device model standards, and are causing manufacturers to build proprietary models with EDA tools that adhere to the new standards.

The behavior of interconnect and device models has changed, due to proximity and material effects. In regards to interconnect, the materials change caused stress on EDA tools because the connectivity and/or the coupling noise was affected by conductors nearby. The material change has also impacted interconnect models because it changed the thickness of wires based on density. This has also had an impact on conductivity and coupling capacitance.

As process technologies evolved to 90-nanometer geometries, proximity effects have begun to affect device models as well. Leakage in the transistors increased, causing library providers to create different kinds of libraries with different leakage/performance trade-offs. This also causes foundries to supply device models that conformed to new standards, some of which are proprietary. In response to these changing factors, EDA tools now need the flexibility to handle new behaviors uncovered in a rapidly-changing environment.

6.2.2 Cell Libraries

Cell libraries are building blocks that designers use while assembling their design. They provide a higher level of abstraction than transistors, enabling designers to build larger designs more efficiently.

Designers use cells, which are design elements from the libraries, to build their design. EDA tools are used to assist in the selection of appropriate cells, and their placement into the design.

Companies such as ARM, Virtual Silicon and Virage Logic provide these cell libraries. In some cases silicon foundries also provide libraries.

6.2.3 Process Design Kits

PDKs are a representation of the process that designers use while constructing their design. They consist of library elements known as p-cells and symbols, as well as other components such as device models, interconnect models, layout technology files and DRC/LVS rule decks. PDKs are available from silicon manufacturers and EDA companies.

6.2.4 Reference Flows

Reference flows provide a methodology for design creation consisting of a flow that streamlines the process of translation from a level of abstraction, such as register-transfer-level (RTL), to a completed design in GDSII. A reference flow defines a methodology, uses a variety of EDA tools, and provides the scripts to run these tools in a specific order to efficiently meet the development team's design objectives. These reference flows come from a variety of sources, including EDA companies, silicon manufacturers and IP vendors.

Using a single design flow across the design team is preferable to using multiple flows. This approach enables the team to share design methodologies, and also be able to share other kinds of design IP across different projects. Productivity is improved because any improvement in the design flow can be shared within the team as a whole, allowing the development resources to be leveraged across multiple projects.

6.2.5 Design IP

Design IP and EDA are key components in the customer's design chain. Design IP is a building block that design teams use as a component of the design; EDA tools are used to integrate the design IP into the larger design.

EDA enables design IP suppliers to develop and provide the models of the IP to design teams in their customer base. Design teams can instantiate these IPs into their design and use the models of the IP to verify the design in context.

There are distinct boundaries regarding IP between suppliers and customers in the design chain. The ultimate quality of the design depends on a range of factors, including the quality of the process, the design IP and the design tools. Due to the legal boundaries between the suppliers, there's an increasing need to model the design information in a manner which communicates the intent, while protecting the IP.

A need has emerged for high-quality, secure models that can be exchanged between an IP supplier and the foundry, the customer and the EDA partners. This enables participants in the design and manufacturing process to exchange data on the behavior and characteristics of the particular IP, without revealing the details of the IP itself. Behavior is contained within the model, but the specifics of the IP are hidden from the end-user. This approach enables the design quality to be assessed across the supply chain without requiring visibility into the IP of the other suppliers.

6.2.6 EDA and Foundries

IC foundries and IC packaging suppliers are two additional components in the customer's design chain. EDA companies work closely with both entities to enable designers to harden designs into silicon at the IC foundries, design the package to be compatible with the silicon and provide a scalable, high-volume solution. EDA companies work with IC foundries to assist in building models of the process technology, and also enable designers to use these models to predict the behavior of the circuit in silicon.

EDA also allows the packaging suppliers to model the behavior of the IC in the package. EDA can help the customer analyze various choices and select the appropriate packages. Once the models are available, EDA also helps the designer predict the behavior of the IC in the package without it being physically implemented at the foundry.

The trend towards smaller geometries requires companies in the design chain to work together more closely, and creates a higher level of interdependence. When designers were working in geometries of 180-nanometer and above, the design flow was easier to manage. Foundries could build a model of the process, hand it over to the customer, give it to their EDA partner, and then let the EDA partner build or enhance the tools used in design. This approach enabled the EDA companies to respond to changes in the process technology by creating new tools or modifying existing ones. The designer can then use EDA tools to build designs targeted to that particular process technology.

As the processes have trended to ever-decreasing feature sizes, there's also been a simultaneous contraction in development cycles and product lifecycle. This has heightened the interdependence of the process, including the tools and the design intent, and has created a new urgency for partners to work together on a real-time basis. Across the design chain, including the foundries, the design IP providers, and the EDA companies, it is increasingly important to collaborate earlier in the development of a process technology. From the inception of the design process to final tapeout, this integrated approach shortens the design cycle and increases the probability that the design will work when hardened in silicon.

Interestingly, the need for an integrated approach and its requirement for stronger relationships between the various fabless model players was accurately highlighted by Dr. Santanu Das, CEO of TranSwitch Corporation at a FSA event in June of 2003, as seen in Figure 6.2.

There is a growing need for the EDA companies to be tightly linked to the manufacturing process. Because of shrinking geometries and product cycles, the time between changes in the process and the time when a customer has to use it has decreased. EDA companies need to understand the evolution in process behavior from one generation to the next, to be able to model that behavior, and be able to predict how an IC will behave in a certain process using that model.

6.2.7 Emerging EDA Challenges

There are three key challenges in the sub-90-nanometer technology nodes. The first is low power, because at 90-nanometer and below the leakage current in transistors becomes very significant. At the same time, the market is trending towards handheld wireless devices. Battery life has become a key factor and power design is critical.

The second challenge is yield, because manufacturing issues at the sub-90-nanometer level are significant. When a tapeout is sent to the manufacturer, the data must be carefully manipulated

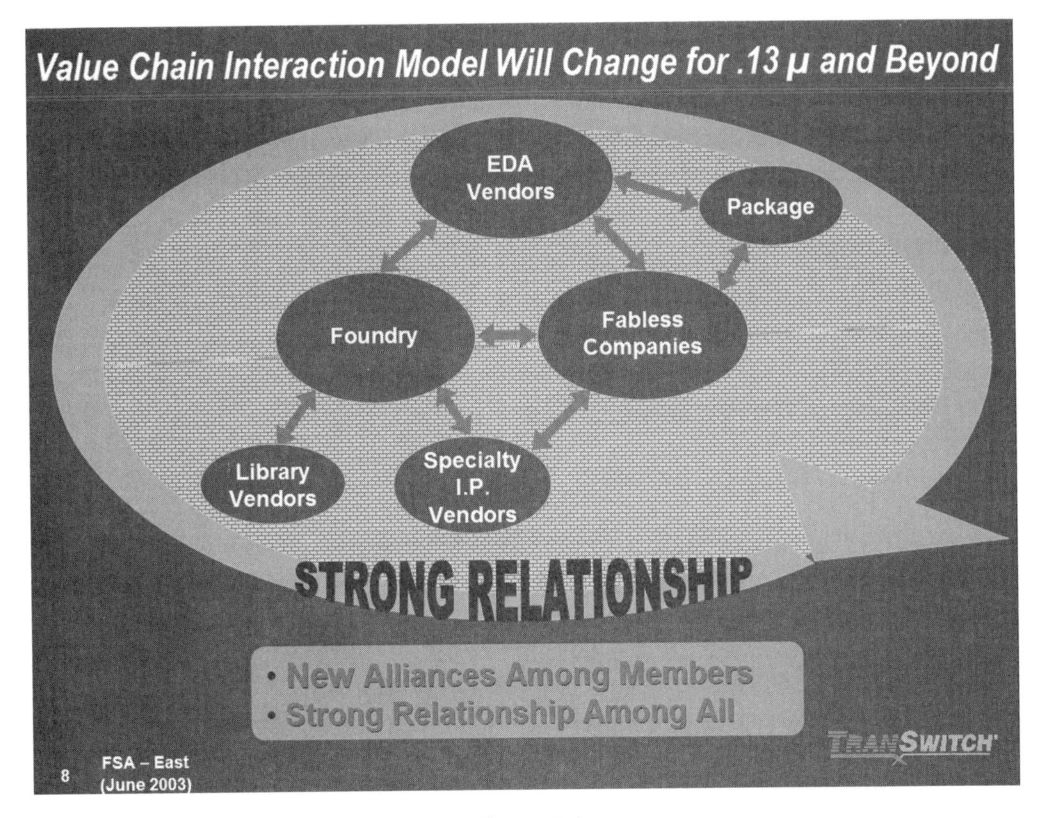

Figure 6.2

to print the circuit in the way that was intended by the designer. The design can become distorted as part of this process, which can negatively impact yield.

The third issue is verification, which is not driven by the size of the IC, but by the forces of Moore's Law. As the complexity and size of the designs continue to increase geometrically, verification challenges increase as well. If a design is not verified in a timely manner, or if the verification is incorrect, costly additional cycles will be required to fix the design. This increases the pressure to produce a viable design the first time, which dramatically increases the chances of receiving working silicon back from the manufacturer in the first pass. To achieve timely and cost-effective development cycles, verification must take place across the entire design process, including the system, RTL, netlist, gate, GDSII, and mask levels.

The increased challenges at advanced technology nodes underscore the need for manufacturers and EDA companies to work more closely to compensate for the impact of the design process on manufacturing. Before the customers tapeout a design, they need to have a predictive model of how the circuit will behave in silicon. By anticipating the effect of the manufacturing process on the design, and by using structured design methodologies, customers will see better yields and a more cost-effective design.

6.3 Physical Design EDA

Nvidia has one of the best physical design flows in the industry. It has a great team, leading-edge tools and yet, even at Nvidia, every tapeout is a major accomplishment. If Nvidia has a hard time managing tapeouts, then what about the rest of the industry? Not every company has all the tools that Nvidia has. Even more significantly, few companies can assemble the breadth and depth of its team.

The practical reality is that fewer and fewer companies can assemble the vast amount of talent required to tapeout at 90-nanometer and below. Does this mean that only a select number of companies are going to be able to take advantage of the speed and density of advanced silicon technologies? The trend is ominous.

6.3.1 *The Physical Design Treadmill*

From a purely technologic point of view, physical design is becoming more and more complex. Professor Kurt Keutzer at the University of California at Berkeley coined the term "physical design triple whammy" to describe the physical design challenge. The growing difficulty comes from three factors (Figure 6.3):

- Smaller transistor sizes turn third-order effects into first-order effects.
- More available transistors drive complexity management.
- Faster clock speeds cause clock skew and noise problems.

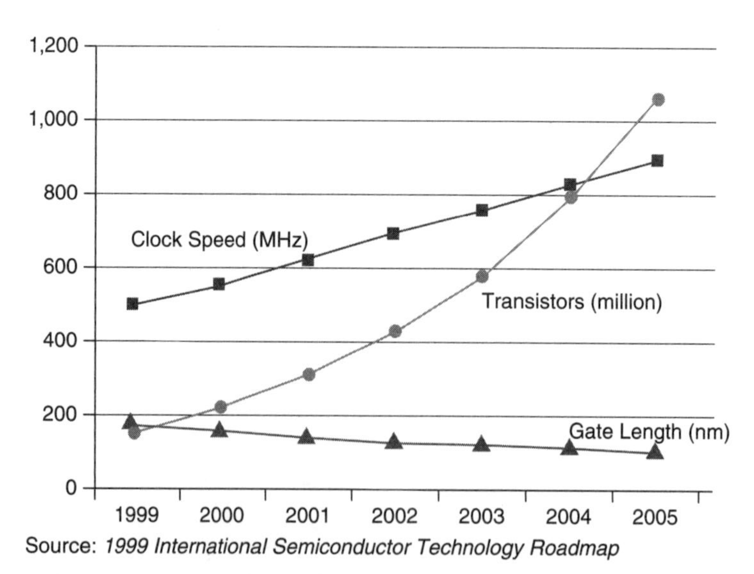

Figure 6.3 Moore's Law is Driving the "Physical Design Triple Whammy"

The downside of having an inadequate physical design capability is becoming more severe. At 0.35-micron, project slips are measured in days or weeks. At 90-nanometer, slips are measured in months or quarters. Major slips can kill a product, and in some cases an entire company. Building a competent physical design capability is expensive. It takes time, tools and engineers. In 2000, Juniper Networks spent $260 million, or $10.8 million per engineer, to purchase Micro Magic, a high-performance chip design house. Most companies do not spend this much for top talent, but the price tag for good engineers (when they can be found) is high. In addition to engineers, it takes a multimillion-dollar investment to purchase the 10 or more design automation tools required for physical design. In addition, it takes 9–12 months to assemble and deploy a working flow.

The bottom line is that the growing cost of staying on the physical design treadmill is draining resources away from product design and development, both of which are critical to the vitality of the fabless semiconductor industry.

6.3.2 The Outsourcing Trend

Many industries respond to growing cost and complexity by disaggregating the value creation chain. As the technical and financial complexity of a value creation step such as wafer fabrication increases, the opportunity for doing a better job through specialization grows. This phenomenon has given rise to the growing outsourcing trend.

Outsourcing is prevalent in the semiconductor industry and has been key to the industry's ability to bring innovative products to market quickly. It started in the late 1960s with the outsourcing of photomasks and assembly and test. It continued in the late 1970s with contract manufacturing and EDA tools. And in the late 1980s, the emergence of pure-play foundries and IP companies began (Figure 6.4).

Outsourced Process	Year Outsourcing Initiated	2000 Market Size (in billions)
Assembly and Test	1968	$8.9
Photomask Reticles	1968	$2.5
Contract Manufacturing	1977	$115.0
Design Tools (EDA)	1978	$3.8
Semiconductor Foundry	1987	$12.9
Source: Custer Consulting Group, Gartner Group, EDAC, Lehman Brothers, and VLSI Research		

Figure 6.4 The Market Size for Electronic Industry Outsourced Design Steps

6.3.3 The Opportunity is Ripe

At first glance it seems that the time is right to outsource physical design. The cost to stay on the physical design technology treadmill is large and growing and the engineering expertise required to play the game is increasingly harder to come by. Just as with wafer fabrication, physical design plays a key role in realizing high-yield, low-cost chips. But this engineering step, as with foundry wafers, adds little value-added product differentiation. Any high-cost step that provides little added value makes for an ideal outsourcing opportunity and physical design (currently performed largely by internal teams) is now ripe for outsourcing.

Unfortunately, it takes experienced engineers to provide physical design services. Therein lies the problem. Any company that wants to provide physical design services needs to attract and retain engineers. Because there are so few engineers to go around, this type of business is extremely hard to scale.

6.3.4 One Path

One time-honored way to address this issue is through the use of automation. Automation has a proven track record for improving engineering productivity. It has also been used very effectively to insulate engineers from complex design issues. In fact, whenever the semiconductor industry has faced productivity issues in the past, it has almost always turned to automation to overcome them. The challenge in this case, however, is huge. It takes 15–20 years of on-the-job training to acquire the know-how to perform state-of-the-art physical design. Software just is not suited to automating these types of tasks. Or is it?

Consider logic synthesis. Logic optimization was performed manually at the gate level 15 years ago, and today it is almost entirely automated. It once took engineers many years of experience to pick up enough insight and tricks-of-the-trade to perform at the top of their field. Now these tricks and many others are available in an automated form to anyone who uses a synthesis tool.

Logic design and physical design are completely different domains, but the potential advantages available from automation are similar. Automation can be used to capture best-in-class practices and make them available to all. Automation can also be used to remove manpower bottlenecks on critical portions of the chip design process.

6.3.5 Summary

The cost and complexity of physical design are growing disproportionably to the value it adds to a final product. At first blush the domain is ripe for outsourcing, but its manpower-intensive nature has made a scalable outsourcing model difficult to realize. Automation has the potential to eliminate manpower constraints and has the added advantage of being able to deploy best-in-class practices across the entire industry. The fabless semiconductor industry will greatly benefit from a software-driven, physical design outsourcing solution.

CHAPTER 7

Intellectual Property

7.1 SIP Industry Overview

Semiconductor intellectual property (SIP or IP[1]) has existed since the advent of the semiconductor industry. In the early years, IC suppliers such as Fairchild, Intel, TI and Motorola developed proprietary SIP (including data and circuit design expertise, process knowledge, packaging test equipment and other items) for their internal use. They fiercely protected their SIP through patents, trade secrets and other legal protections. Occasionally SIP was licensed to third parties, but this tended to be the exception rather than the rule (Figure 7.1).

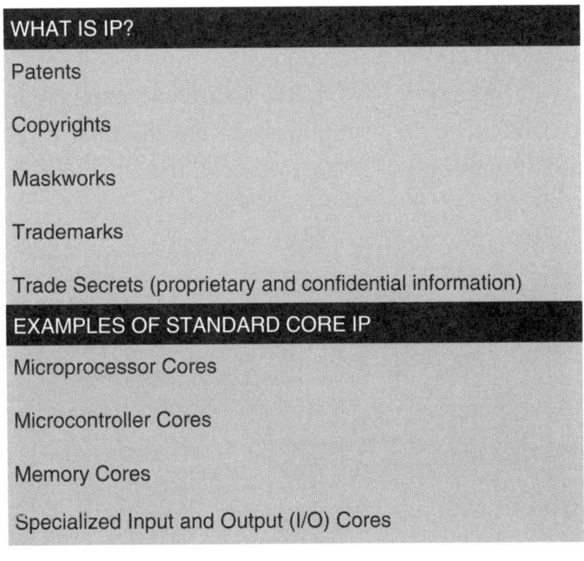

Figure 7.1

Over the last decade, IC design productivity has failed to keep pace with Moore's Law and a "design gap" has emerged. This is despite the development of a $3 billion per year commercial electronic design automation (EDA) industry. In response to this increasing

[1] It should be noted that IP is used throughout this chapter in the industry sense; when referring to the legal concepts of intellectual property the term "IPR" is used.

capacity, IC suppliers began looking for ways to close the gap between what engineers could reasonably be expected to design within a given time frame while still meeting schedules, and the capacity of the silicon (sometimes referred to as the "design productivity gap"). They began to develop reusable SIP that contained increasingly complex functionality. As the foundry model emerged, fabless IC companies adopted a customer-owned tooling (COT) design flow and a block-based design methodology partly to leverage the cost advantages of the foundry model, but also to allow mobility of design. Many of them developed their own SIP but outsourced when possible.

Perhaps the best example of the emergence of SIP as a discrete entity licensed to third parties is the ARM core from ARM Holdings PLC. ARM developed a RISC processor architecture initially for PCs; however, it saw a broader opportunity and began to offer an embedded processor core that could be used in application-specific integrated circuit (ASIC) design. ARM's product is available in both a soft (synthesizable) and hard version. Since the ASIC design flow many companies use relied on the use of "libraries" of standard cell, memories and I/Os as building blocks, a number of "library" suppliers emerged, notably Artisan Components (since acquired by ARM). The high content of embedded memory in ASICs was also addressed by companies like Virage Logic. These IP blocks are used today by several pure-play foundries, including Taiwan Semiconductor Manufacturing Corporation (TSMC), United Microelectronics Corporation (UMC), and Chartered, and also by many integrated device manufacturers (IDMs). The net result has been the emergence of a commercial SIP industry that grew from $757 million in 2002 to $2.1 billion in 2006 (Figure 7.2).

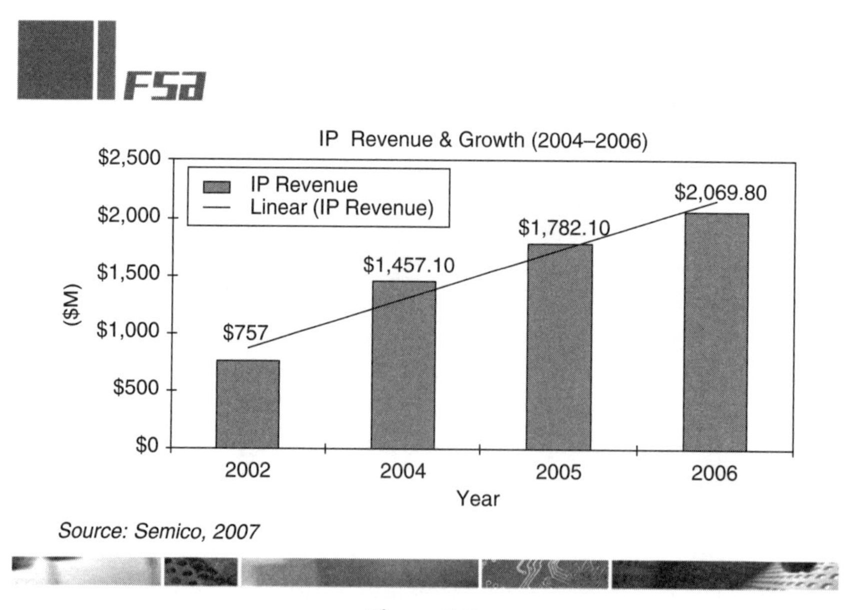

Figure 7.2

SIP is critical for the design and implementation of complex system ICs. As designs become more complex, a greater number of SIP products are being embedded in IC designs. SIP has become a key part of the electronic design process because it can reduce IC development costs, accelerate time-to-market, reduce time-to-volume and increase end-product value; in short, it can provide a solution that enables companies to bridge the "design gap."

However, in some ways the SIP commercial business infrastructure is like the western United States in the mid to late nineteenth century: wild, unpredictable and untamed. Providers have introduced a variety of business models intended to capture value, maximize revenue and accommodate customer needs. A major portion of the industry still consists of smaller private companies many of whom also provide design services. With the increasing use of commercial SIP from multiple sources and the wide variety of SIP business models, the process of finding, evaluating and purchasing SIP has become time-consuming and more complex than it should be.

7.2 SIP Business Environment

SIP business practices include elements similar to those found in the traditional semiconductor or ASIC, EDA and design services markets. However, unlike the well-established business models in the ASIC and EDA industries, SIP business models tend to be more complex because several parties in the supply chain are involved with the successful deployment of commercial SIP in an IC design. For example, an IC that incorporates multiple SIP products may involve several SIP suppliers and more than one source of manufacturing. As a result, buyers may need to manage multiple vendors in the supply chain, each with different business models and technical capabilities.

While some uniformity in business practices and SIP business models has emerged, the industry has not yet standardized due to the wide variety of SIP product types, customer needs and frequent changes in EDA tools and process technologies. SIP products are generally available in two forms: soft and hard. In addition, there are several classifications of SIP products such as, foundation, processor, memory, library, analog/mixed-signal and connectivity SIP. Other classifications of SIP such as verification or software SIP may also be included in the above products, or marketed separately as stand-alone products. The form and type of SIP products can influence the price of a license to the SIP product, how readily available it is, how often it is updated, the type of support it may require, and how quickly it can be integrated in the end product. The value of an SIP product also varies based on customer utilization.

The following description provides a basic understanding of the forms and types of SIP, and, at the same time appreciate the complexity of the industry.

- *Soft SIP*: This is usually offered in high-level language (usually RTL, C++, Verilog or VHDL) or sometimes in netlist format. Soft SIP is generally portable (it can be moved from one foundry process to another without great difficulty), but it is not optimized for a specific process technology. Thus, the power, performance and area are not known until it is committed to a specific process technology and library. The value of the SIP is typically its functionality, reusability, availability and possible conformance to an industry standard specification such as Universal Serial Bus (USB) 1.0, ARM AMBA 1.0, PCI, or IEEE 802.11. Some suppliers of soft SIP may also provide netlist or programming data that would allow implementing the SIP in a FPGA for validation or in a finished product.

- *Hard SIP*: Hard SIP is usually offered in GDSII format along with an accompanying set of EDA views or models, and is optimized for a specific foundry process. It may also be offered in "bit-stream" format if the originating soft SIP was "hardened" for a particular FPGA device. Hard SIP often has a data sheet detailing items such as power, speed and area similar to that found with a finished discrete IC. Examples of hard SIP include processors, standard cells, memories, phase locked loops (PLLs), I/Os and analog blocks. Hard SIP is generally not portable to another foundry process and even though portable libraries exist (usually consisting of standard cells, I/Os and memory) they are generally not optimized for any specific foundry. Some suppliers of hard SIP conduct extensive silicon validation on test chips to ensure high quality and yield.

- *Foundation SIP*: This often includes standard cell libraries, PLLs, I/Os and small macros. It is readily available mostly as hard SIP and is usually offered as a single package; that is, several items of foundation SIP are licensed to the end-user as a single package and not as discrete SIP products. General-purpose foundation SIP may be offered at no cost by suppliers and foundries to attract customers. Specialty foundation SIP (i.e. low-power or high-performance) is typically priced based on specific customer needs, level of customization and process technology supported.

- *Memory SIP*: Memory SIP is generally provided as hard SIP targeted at a specific process node, sometimes together with memory compilers. A wide variety of memory types and specifications are often needed on a single system IC. This can include SRAM, non-volatile (Flash, EEPROM), DRAM and 1T-SRAM. Higher density memories may be offered with redundancy to improve yield, error detection and correction, and even integrated test and repair. SRAM types may include single- and dual-port (1rw, 1r/1w, 2rw) register files and SRAMs; multi-ports; content addressable memory (CAMs) and custom designs that may be further optimized for area, power or speed. Memories can be generated with memory compiler tools created with foundry and process-specific bit cells or on an individual basis as macros. Memory compilers enable users to customize specific features such as aspect ratio, power bussing, redundancy, etc. On the 0.13-micron process node, as many as 25 different memory compilers may be needed to support designs. Since memory uses special design rules that allow for higher density, it is a "defect magnet." It is often used to drive silicon processes so validation is a key requirement.

- *Connectivity SIP*: This is also known as standards SIP; it implements a given specification or standard such as USB, PCI, IEEE1394 (commonly known as "Firewire") IrDA, Bluetooth or 802.11. Connectivity SIP may be available in soft or hard form. Its availability is often closely related to a specific revision of the standard it supports, which may be continuously evolving. While "connectivity" generally refers to off-chip buses or interfaces, it could include on-chip buses such as AMBA or OCP.
- *Processor SIP*: Processor SIP includes microprocessors, digital signal processing (DSPs), MPEG and others. It is typically available in soft form and may be hardened to support any process technology. Some foundries offer hardened processors for companies that do not already have a license or want to know what specifications will be at the time of selection. When referred to as "silicon-proven," this means only for selected process technologies. Processor SIP tends to be of the highest value because of its rich architecture, software and hardware support infrastructure.
- *Analog SIP*: This includes a variety of functions such as digital-to-analog converters (DACs) or analog-to-digital converters (ADCs), crystals, voltage detectors and other items. It is always provided in hard form and for a specific manufacturing process technology. It is highly sensitive to process changes and may require re-design if these changes occur during IC implementation.
- *Platform SIP*: Platform SIP is an emerging segment of the industry focusing on products that enable development of a "structured ASIC," application-specific standard products (ASSP) or platform. This may include system-level integration functions and some programmability (modular array ASIC, metal programmable logic, FPGA cores, and non-volatile memory) that allows the platform to be used by multiple companies. Platform SIP can include bus interconnect architectures, power, clock and other physical structures.

Industry-standard data deliverables for these cores are in Verilog, GDSII, SPICE, EDIF and ASCII formats. The Verilog register-transfer-level (RTL) model is also referred to as the "soft core." The desired characteristics are implemented by system chip designers through the selection of appropriate standard cells used during the re-targeting process. Field programmable gate arrays (FPGA) versions are plug replaceable for the standard product hard core and have been validated in the standard product test bench. When the soft core is floor-planned, synthesized, placed and routed using an ASIC standard cell library, the soft core becomes a firm core based on the RTL model. A manually optimized core in GDSII graphical data format is thought of as the most optimized and fixed; it is referred to as a hard core.

Further adding to the complexity of SIP business models is the variety of distribution channels, each of which may influence how the SIP is licensed, maintained and supported. For example, certain foundation SIP may be downloaded from the Web site of the provider or foundry, though most commercial processor and memory SIP products are licensed through a direct sales channel and require longer technical evaluations and more complex licensing agreements. Most specialty SIP (cell libraries, I/Os and analog SIP) may require close interaction with foundries (some of which offer their own specialty SIP directly, through providers or through EDA vendors).

In addition, it is common in the industry to develop custom-built SIP through a service contract and subsequently offer it as stand-alone product when it matures. Once the SIP matures, it may be classified as a "commercial" product that is offered with technical support, including maintenance updates when the foundry design rules change or when re-characterization is required due to process changes.

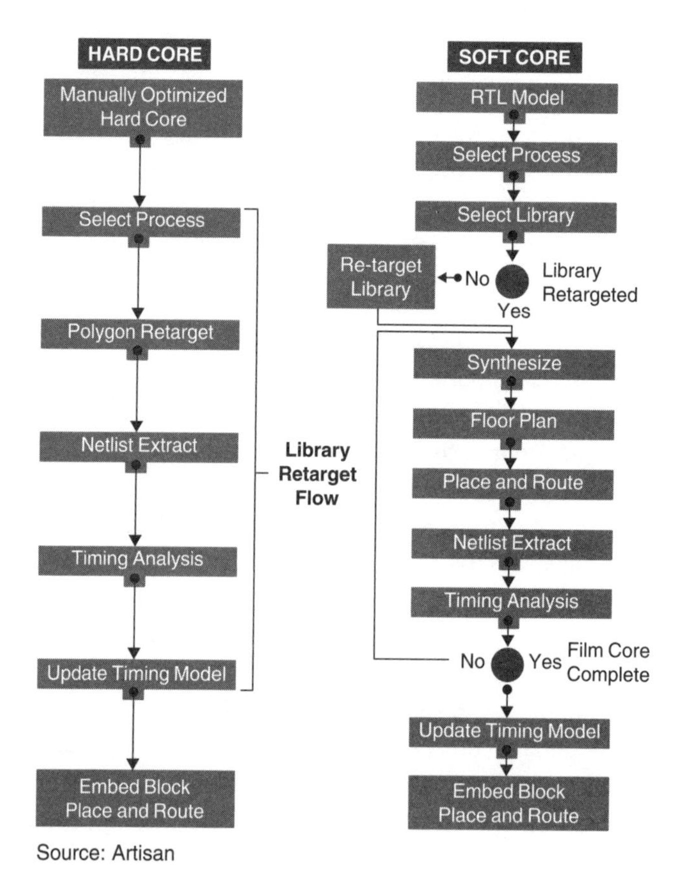

Figure 7.3

7.2.1 Hard and Soft Core Re-Targeting

The design flow for re-targeting a hard core and a soft core model is shown in Fig. 7.3. The hard core steps include: (1) select the desired process; (2) re-target the GDSII layers;

(3) extract the SPICE netlist; (4) simulate for new timing and (5) update the timing model. The core is ready to be embedded in the desired application. Not only are there half as many steps required with the hard core, there also is significantly less effort per step since the GDSII data base was originally proven in silicon. With the block place and route system for system-chip design, the hard core is thought of as a standard cell and the design flow follows the standard cell re-targeting flow. The design flow to re-target a soft core is more complicated. The soft core steps include: (1) select the process; (2) select the library; (3) re-target the library if it has not been re-targeted; (4) synthesize with the library; (5) floor plan for placing and routing; (6) place and route the standard cells; (7) extract a netlist for verification; (8) simulate to ensure the desired timing is achieved and (9) update the timing model.

If, at this time, the firm core does not meet the desired criteria, then repeat steps 4–9. The firm core is now ready to embed as a hard core. The RTL firm core can be re-targeted using the same steps as the manually optimized hard core after it has been optimized.

7.3 Sourcing SIP Products

Deciding whether to develop internally or source commercial SIP products depends on several factors such as whether or not the SIP product meets the specific project requirements, availability, specification, cost, number of sources, the reputation of suppliers and number of foundries supported. Outsourcing is unlikely if the SIP product is seen by the potential buyer as a core competency or a key differentiator in his product, or if its use requires third-party access to the buyer's patents or trade secrets. To simplify the process of finding, evaluating and purchasing SIP products, these factors should be considered as a starting point in the planning stages of a design project.

- *Sourcing*: Sourcing the SIP product can be as simple as browsing Internet-based SIP product catalogs or Web sites from providers and foundries. The difficulty is usually in determining the suitability of an SIP product for a given application. In the early stages of the design and during the system-level architecture phase, the primary focus tends to be on high-level requirements, the types of functions needed, where to get the SIP Product, how available it is and what it costs. The manufacturing source is often overlooked or seen as a secondary concern, unless the buyer has already chosen the foundry. However, the manufacturing source should be considered up-front and together with the choice of the SIP product, as it can affect availability, cost and specifications, particularly for hard SIP products.

- *Evaluating*: Evaluating the SIP product to determine whether it is technically and commercially feasible to instantiate an SIP product in a given design is a more complex process. It requires knowing the variety of SIP product types and forms, maturity, process mobility and business models involved. It may be very hard to perform a "like for like" comparison of the economic value for two similar different

SIP products. Some key factors that potential buyers of SIP products should consider are:

- *SIP Form*: If the SIP product is in soft form, then buyers should consider what it would take to prove the SIP product in silicon, and factor in the cost of physical design on top of the SIP product's price. If the SIP product is in hard form, then buyers should consider how mobile it is and factor in the cost and risk of changing foundries, if need arises.
- *SIP Type*: If buyers are looking to source a connectivity SIP product, then they should consider how frequently the standard is changing, whether the SIP product complies with the standard, and what updates the provider will offer if the standard changes. If buyers are looking to source a processor SIP product, or any other complex functional SIP product, then they should evaluate whether its architecture and software tools meet their specific needs, and factor in the cost of any modifications needed. If buyers are looking to source specialty or analog SIP products, then they should consider its silicon performance and how it may be affected by process variations.
- *SIP Mobility*: If buyers are looking to source a hard SIP product, then they should consider how mobile or how portable it is. This is a key factor that should be evaluated up-front with the manufacturing source. Foundry-provided SIP products are usually not technically portable, or their use in another foundry process may be restricted by license. This presents a potential time-to-market risk during shortages in foundry production capacity. Buyers should carefully weigh the pros and cons of sourcing SIP products that cannot be moved. Alternatively, buyers can look for multiple sources of SIP products they might substitute if replacement becomes necessary.

- *Licensing*: Licensing the SIP is the final step in the sourcing process. This step can be very time consuming, considering the variety of business models and licensing agreements that can vary based on the buyer's needs, SIP product form, type and process technology. In addition, if more than two parties are involved, licensing transactions may be more complex as there may be additional considerations relating to cost, mobility and sublicensing.

7.4 Baseline Terminology

Even though some of the SIP products business models originated in EDA, SIP products can be very different from EDA products in how they are licensed, maintained and supported. In general, soft SIP products tend to bear more resemblance to EDA – and indeed to the software industry in general – than do hard SIP products. Hard SIP products tend to bear more resemblance to discrete ICs and are treated very differently. To avoid any confusion it is necessary to define some key terms that are used through the remainder of this chapter:

- *SIP Product*: The term "SIP product" means the design files, models characterization data, test benches and other items that make up the product offered by providers to buyers. An SIP product is a "virtual good" that is licensed to one or more buyers while ownership of the underlying IP rights remains with the provider. SIP products have detailed functional specifications and – if hardened – well-defined operating characteristics, all of which are essential for integration into the end device, as well as achieving its intended functional specifications.
- *SIP Purchase*: The term "SIP purchase" means the non-exclusive transfer of a limited subset of the provider's intellectual property rights (IPR) in an SIP product to a buyer. With an SIP purchase there is normally no wholesale transfer of ownership of the IPR in the SIP, rather there is the granting of a license to some limited IPR from the provider to the buyer. An SIP purchase is usually accomplished through an SIP License Agreement.
- *SIP License Agreement*: The term "SIP License Agreement" means the contract that defines the terms of the SIP purchase; more simply it defines how, where and when the buyer may use the SIP. This typically includes the right to manufacture ICs containing instantiation(s) of the SIP and may limit the use of SIP to a pre-defined field.
- *SIP Rights*: The term "SIP rights," means the scope of permissible uses for the SIP defined in the SIP License Agreement. The SIP rights may vary depending upon the SIP form, type, maturity, provider's business objectives and buyer's needs. SIP rights are broken down into five basic categories, use, copying, modification, distribution and sublicensing.

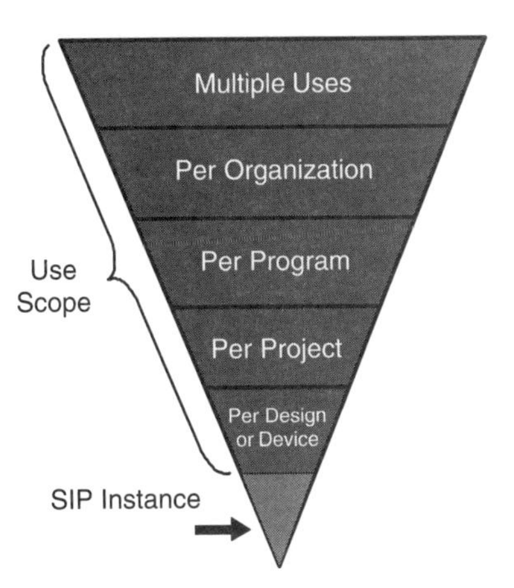

- *SIP Instance*: The term "SIP instance" means a specific configuration of the SIP that can be used one or multiple times as defined in the SIP rights "use" category.

- *SIP Use Scope*: The term "SIP use scope" means a set of restrictions that are determined by the business model and the license agreement. SIP use scope is generally broken down into the categories described in Figure 7.4.

Per Design	An IC intended to be manufactured and distributed commercially
Per Project	One or multiple chip designs for a single application segment
Per Program	One or multiple projects for a variety of application segments
Per Organization	One or more geographies (incl. corporate-wide) and/or programs
Multiple Uses	Many uses over a period of time for given process technologies
Source: FSA	

Figure 7.4 SIP Use Scope

SIP use scope can range from narrow to broad, which suggests that – like EDA licenses that are tracked by the number of CPUs or number of users – SIP is tracked by its usage and restrictions that are intimately related to the SIP value.

- *SIP Tracking*: The term "SIP tracking" refers to the technical measures implemented by providers to identify, track and monitor usage of their products. A typical mechanism for tracking a hard SIP product is to assign special layers in the SIP Product used for identification purposes and to restrict buyers by license from modifying these layers.

7.5 Finding SIP and Related Products

Direct Sources: Buyers can locate SIP products, related tools and design services directly through the Web sites of providers, foundries, EDA vendors and broad line IC suppliers.

Internet Catalog Listings: A popular method for locating SIP products is though on line Internet catalog listings, such as Chip Estimate and Design & Reuse (D&R). Both sites include a comprehensive listing of SIP organized by provider, type, application segment, etc. as well as search engines to find the SIP.

7.6 Evaluating SIP Business Models

7.6.1 Introduction

Buyers who deal with multiple sources for SIP products may have difficulty coping with the variety of business models, and the different fees involved. Often, it is very difficult for buyers to compare the value of similar – or even functionally identical – commercial SIP products. Unlike the ASIC and EDA marketplaces that have mature and relatively consistent business models, the SIP business infrastructure can be confusing. Buyers often feel that it takes a significant effort to decide whether to design or source SIP products from third parties because it is tedious to determine the economic value of different SIP products.

Without any objective means to compare differing business models, some of the value of design reuse may be lost. Two identical SIP products purchased on different business terms may have different value when analyzed in light of the buyer's specific needs. As a result, what is needed first is an objective taxonomy of the prevalent business models in the current SIP product marketplace. By using this taxonomy, buyers can sort out the economic elements among different business models and make informed decisions on whether to design or source SIP products.

The following section contains a description of the prevalent business models in the current SIP product marketplace, what is typical to expect from different providers and considerations for determining the economic value of different SIP product types. The discussion of each business model includes:

- The purpose of the particular business model
- A definition of the payment options
- Typical structure of the fees paid
- Most common SIP use scope for the SIP products

Figure 7.5 provides a summary of the principal attributes of business models for established providers and SIP products. While this may appear to be a simplification, in reality a typical SIP purchase may involve elements of more than one business model.

	Per-Use	Time-Based	Royalty-Based	Access
Purpose	Fee for each SIP on defined use scope	Multiple uses of SIP over a period of time	Amortize cost of SIP Share risk-reward	Fee for SIP portfolio over a period of time
Payments	Event based	Time-based	Value based	Subscription based
Structure	One time fee for a design (first or subsequent)	Fee for all designs within a given time	Some or all fees spread across units	Up front fee plus discounted use fee
Scope	Per design Per device	Multiple uses Per device	% of Unit value Per device	Multiple SIPs Per organization

Source: FSA

Figure 7.5 Typical Business Models for SIP Products

Another aspect of determining the economic value of an SIP product is related to the different fees for enabling the successful use of the SIP product.

Figure 7.6 provides the typical enabling components that usually represent a secondary revenue stream for providers. These are often structured as separate fees within the SIP License Agreement itself, or sometimes as a separate Statement of Work (SOW) or contract if there are specific needs that are non-standard. These components enable:

- The use of the SIP product itself
- The IC design in which the SIP product is instantiated
- Specific needs relating to SIP product use and deployment

Per-Use	Maintenance	Support	NRE	Contract Service
Purpose	SIP updates, bug fixes, and revisions	Address specific customer needs	Enable SIP use	Enable IC design
Payments	% List price License fee	Scope based	Milestone based	Hourly based
Structure	Part of initial license agreement (May be induded)	Basic package or, separate contract	Initial fee % milestones	Initial fee (SOW) % milestones
Scope	Changes in spec, process tech, etc.	Web, email, tel, on-site, geography	Modifications, re-spins, porting, etc.	Tool runs, IC Integ, EDA views, etc.
Source: FSA				

Figure 7.6 Typical Enabling Components for SIP Products

Use of consistent terminology when discussing SIP product business models and their enabling components is essential in the industry, particularly where SIP purchases from multiple suppliers are involved. Consistent terminology can simplify negotiation, provide a consistency in the structure of payments, and align expectations between two parties in the SIP purchase.

7.6.2 SIP Business Models

7.6.2.1 Per-use Model

In the per-use model, the SIP purchase gives the buyer the right to use each SIP product within a defined SIP use scope. The per-use model is common where the SIP product requirements and usage are well understood early in the design phase. The fee paid may apply to a specific SIP product instance (such as a single use of a DSP core) or an SIP configuration containing several instantiations of the SIP product that may be treated as a single use (such as instantiating multiple SRAMs in a single IC design). Since what constitutes a "use" is often the subject of disagreement, it is important for providers and buyers to agree early which "uses" trigger payment obligations and which do not.

The payments in a per-use model typically include an initial fee for the first use normally paid upon execution of the SIP License Agreement. If the SIP product requires some development work to meet the needs of the buyer's application, then the initial fee may be a percentage of the license fee with subsequent portions being paid upon a milestone or an acceptance event (such as the delivery of EDA views or GDSII files to the buyer).

The per-use model often includes a subsequent discounted reuse fee that is paid for a new SIP use scope. If there is a reuse fee, it is typically paid when the end product goes to production and not for prototype runs or bug fixes.

Other fees may be paid for derivative uses of the SIP product; these may include modifications, re-spins or new configurations of the SIP product. Derivative uses are generally treated as a non-recurring engineering (NRE) and often subject to a new contract, but some providers may treat such derivative uses as an entirely new product with a new SIP use scope.

The SIP use scope in a per-use model is usually per device, or per project (which may include one or multiple chips), and the right to manufacture is normally restricted to a specific process technology.

7.6.2.2 Time-Based Model

In the time-based model, the SIP purchase gives the buyer the right to use the SIP multiple times over a defined time period. The time period may be explicit, such as a fixed date, or implicit in that the use of SIP may be restricted to a specific process technology that will become obsolete over time.

The time-based model for SIP products is different from the time-based license (TBL) model used in the EDA industry. In a TBL, the user licenses the right to use a tool and the IP in the work product of that tool belongs to the user. By contrast, the time-based model for SIP products usually provides the buyer with the right to design and manufacture with the SIP. In most SIP time-based models, there is no transfer of the underlying IP rights in the SIP product. In addition, maintenance and support for SIP products is different from EDA tools; this will be discussed later.

The payments in a time-based model include an initial fee paid on signing of the SIP License Agreement. This fee normally covers multiple uses during a specified period of time or limited to a specific process technology node. The SIP License Agreement may expire at the end of the time period and/or end of life of the specified process technology, although it may be renewed on expiration or sometimes if the buyer adopts a new foundry or process technology.

Some SIP products may be offered in both per-use and time-based models. In such a case, the buyer should expect the pricing of the time-based model fee to be some multiple of the per-use fee. For example, the fee for a perpetual license of a foundation SIP product may be between three and five times the amounts for a per-use model.

In some SIP products that include companion software there may be a TBL of the software that is separate from the license for the SIP product. There are two types of companion software: generators and embedded software. Examples of generators include memory compilers, programmable processors (parameterized SIP) and PLL compilers. Examples of embedded software – which may also be offered as stand alone software IP – include application layers, device drivers, protocol stacks and real-time operating systems (RTOS).

The SIP use scope in the time-based model is generally unlimited. Buyers can use the SIP product as many times and in as many projects as they wish within the time limits of the SIP License Agreement.

7.6.2.3 Royalty-Based Model

An increasingly common model in the SIP products industry is the royalty-based model. Here, the provider and the buyer agree to amortize a portion of the license fee over the life cycle of the end product and thereby share the risks and rewards of the SIP product's use. The royalty-based model may be appropriate for buyers who wish to minimize their up-front costs and SIP providers who are willing to accept a lesser amount up-front in return for potentially greater long-term rewards if the buyer's end product is successful.

In the royalty-based model, there is an ongoing payment of consideration tied to some objectively quantifiable measure, such as the number of units sold or the number of wafers manufactured. The royalty-based model is most common in the licensing of highly differentiated SIP products and may include other parties to the agreement such as the foundry who manufactures the device instantiating the SIP. In some cases, royalties are paid by foundries, such as in the case of foundation SIP products; in other cases, royalties are paid by the end-user such as in the case of processor SIP products and specialty memory SIP products.

Payments in the royalty-based model typically include an initial fee paid upon execution of the SIP License Agreement, followed by subsequent per unit fees based on some measurable criteria agreed up-front between the provider and the buyer. In certain cases the initial fee may be discounted, waived or treated as an advance recoupable against ongoing royalties. Per unit fees usually apply to the total number of production units manufactured. Royalties are normally due and payable within 30 days of the end of the calendar quarter in which the units are shipped, or in which the defined criteria have been met. Providers normally include audit provisions in the agreements to ensure timely and accurate royalty payments.

The SIP use scope in the royalty-based model is normally on a per unit basis, the "unit" being the device in which the SIP product is instantiated, typically the finished end product or the wafer. For most SIP types, payments are structured as a percentage of the manufactured cost or the average sales price (ASP) of the end product. When the SIP product is replacing a discrete device such as analog or memory, payments may be structured as a flat ASP per unit calculated as a percent of the price of the equivalent discrete IC minus packaging, manufacturing and testing costs. For other SIP products, payments may be structured as a percentage of the measurable value the SIP product provides to the end product. For example, a royalty may be structured as a percentage of the cost savings due to die size or yield improvement; or a percentage of the end product price premium gained due to performance improvement.

It may often be the case that additional royalties have to be paid to a third party. It is important for buyers considering a royalty model to be aware whether such royalties apply. Failure to do this may result in unexpected cost and delay in the use of the SIP product. Some of these payments may be part of a well-defined royalty or patent pool, such as the MPEG or IEEE1394 that normally requires payment of a royalty by the end equipment manufacturer; others may be less well defined and might require additional payments of different amounts to one or more participants in the pooling arrangement. These third-party royalty scenarios all differ in some respects, so it is important for buyers to look into whether they are facing such payments for use of a particular SIP product before committing to the use of an SIP product.

7.6.2.4 Access Model

In the access model, the buyer generally is granted access to an SIP product portfolio for a defined period. This model may be useful when the buyer does not know his design needs up-front. It may also be attractive for design service groups that are seeking to widen the range of market application segments they support and thereby attract more design projects. The advantage of the access model is that it allows the buyer lower cost access to a wide range of SIP products that may be used as and when needed.

The payment structure for the access model generally includes some sort of up-front fee paid upon execution of the SIP License Agreement. This fee enables the buyer to design with any of the SIP products and in some cases may include the right to manufacture products using the SIP products. There may also be a subsequent annual fee that is sometimes discounted based on usage. In some cases, there may be an additional per-use fee paid upon production, if the SIP product is used in a manufactured device.

The actual amount of the access fee varies based on a number of factors such as the breadth of the SIP product portfolio accessed, the target application segments and the buyer's specific support requirements. In some cases, providers may provide a design service company with access to an SIP product portfolio at no charge in exchange for a license fee paid by the end customer. In other cases, providers may discount the access fee for design aggregators such as IDMs or fabless ASIC vendors, in exchange for alternative revenue streams, such as tapeout fees or royalties.

The SIP use scope for the access model can be broad and may include multiple uses-per-project, site or location, or be unlimited across the buyer corporation and its subsidiaries. The right to manufacture may include the right to use multiple foundries and process technologies.

7.7 SIP Product Enablers

As mentioned previously, the license fees represent a primary revenue stream for established providers. Another aspect of the SIP products business is the customer's ability to successfully deploy and deliver end products instantiating the SIP product. As a result, it is common for providers to offer additional services. Because these services and associated fees enable either the SIP or the end product, they are termed SIP enablers.

7.7.1 Maintenance

Before discussing the particulars of maintenance, it is important to draw a distinction between warranty, maintenance and the support of an SIP product. Warranty repair is discussed in a later section. Maintenance is related to the SIP product's specifications and applies only to the SIP product as a discrete entity. Support, on the other hand, is primarily related to the specific customer needs, which may be intimately linked to the SIP use scope.

Maintenance for soft SIP products is similar to the concept in the EDA business. It covers the provision of minor updates, bug fixes and any other revisions to the SIP product because of a change in functionality or specifications. Maintenance for hard SIP products is much more resource-intensive, since it must include the costs of manufacturing and testing enough devices to ensure conformance to specifications and achieve silicon performance. Since the quality and performance of SIP products depends on numerous external factors such as changes in standards, process technologies and EDA tools, maintenance is a major cost component of the SIP products business. This may be amplified by the fact that maintenance costs may need to be amortized among a smaller number of users.

Maintenance fees are calculated based on the list price of the initial SIP product and they usually range from 12 to 18 percent per annum. Unlike the SIP license fee that is sometimes discounted, maintenance is usually not discounted.

Maintenance may be provided under a separate agreement whose term might extend beyond the term of the SIP License Agreement. Some providers may occasionally bundle the cost of maintenance into the license fee or require that the buyer take a maintenance contract that runs with the term of the SIP License Agreement. In addition, they may also include basic support such as e-mail, telephone and on-site support as part of the maintenance offering. However, this does not reflect the norm of the industry and it is likely to phase out as the SIP Products industry matures.

It is essential for buyers to understand what services are provided under maintenance, and what services are not. To ensure that there are no future surprises, buyers should consider the following factors for each SIP product they seek to license:

- *Changes in Functionality and/or Specifications*: It is important to know which updates are covered and which are not. What are the functional corner cases and the test bench coverage? What changes are needed to adhere to changes in the specification and what is covered under warranty?
- *Process Technology*: The integrator should know which and how many updates are covered for the licensed process node. How will process changes affect the SIP product's silicon performance? How portable is the SIP product and what mobility alternatives does the provider offer?
- *Connectivity Standards*: Define up-front which revision of the standard is supported. How is compliance with the standard ensured? How does the provider handle future standard changes?
- *EDA Tools and Formats*: Determine which EDA tools, models and what interface formats are supported. Which tool version is supported? How often are the EDA tools views and models updated? What is provided as part of maintenance?
- *End-Product Integration*: Determine how the SIP product performs in the context of the IC. What are the corner cases? What services are offered to enable the IC using the SIP Product?

Most providers will typically specify what non-conformities are covered and how many updates are included to cover changes in process rules, EDA tools and future revisions of standards. For example, certain process corners may be not covered, or there may be limit on process changes may be covered; or for some types of SIP products and leading-edge process nodes, any new characterizations may be treated as NRE and not covered under the maintenance agreement.

Buyers should insist on clear, unambiguous statements of how often updates are made available, response times for bug reports and the provider's stated time commitments for fixes to critical, serious and minor bugs.

7.7.2 Support

Because the SIP use scope varies by business model, SIP product type and customer needs, the support requirements for different customers may vary significantly. As a result, it is difficult to generalize as to what constitutes "standard" support for SIP products. Even though most providers offer a pre-defined basic support package, support is often negotiated separately and is in a separate contract.

The pre-defined basic support package usually includes e-mail support, some kind of Web-based "helpdesk" and possibly phone access to application engineering staff. Others may provide some basic level of on-site application engineering support and/or some measure of R&D support.

Often providers may also offer tiered or custom support packages that extend the level of support based on the buyer's needs and the demands such support places on the provider's engineering resources. These packages are priced based on the level of support needed by the buyer and the resource commitment required by the provider; the number of users, simultaneous projects and multiple geographies that must be served may complicate this pricing. These support packages may specify a pre-defined level of support by the number of engineers, simultaneous projects or geographies served. They may also include training, expert support for SIP product modification and expert support on the end application.

Buyers should make sure that the support contract clearly identifies which services are covered and which are not as part of a standard support package. In addition, they should anticipate the future levels of SIP product usage and expertise requirements and work with the provider to determine an appropriate level of custom support, if needed.

7.7.3 NRE Charges

NRE fees are typically paid to a provider to adapt, modify or optimize an existing SIP product, or sometimes to develop new custom SIP products designed for the buyer's specific needs. As such, NRE fees are paid to enable the use of the SIP product: the engineering work is central to the functionality of the SIP product and not necessarily the end application.

With an NRE contract there is typically an initial fee paid upon execution of the agreement followed by subsequent fees paid on a per milestone or event basis. Fees tend to vary based on the agreed scope of ownership of the IPR in the resulting SIP product(s). Providers may charge a lower fee and grant buyers a limited license, but if the buyer wishes to own all of the IP rights in the SIP product then the fee will be significantly higher. In some rare cases, the newly developed SIP may result in some form of joint ownership where multiple SIP products are merged into a new one; this may complicate business transactions if both parties plan to offer the SIP product commercially.

Some providers believe that any SIP product that comes from modifications is an entirely new product that must be treated as such, with a new SIP License Agreement and a new SIP use scope. This classification of the modified SIP product as a new SIP product is largely due to revenue recognition principles that may favor new products over ongoing NRE. If the SIP is treated as a new product, the provider can recognize most of the revenue upon delivery to the buyer. If, on the other hand, modifications are treated as NRE, then the payments to the provider are normally event-based and could involve special acceptance criteria that can delay recognition of revenue.[2]

The scope of NRE work may vary with the SIP form, type and maturity. Examples of minor NRE work are simple SIP product modifications, custom configurations and ports to a new foundry process. Examples of major NRE work include, analog SIP product development, porting of specialty SIP products into a leading-edge process technology and integration of multiple SIP products into a single platform. The exact nature of the work, the deliverables and the cost and timing are normally outlined in a written SOW.

7.7.4 Contract Services

We define "contract services" as outsourced engineering work such as design services, custom EDA views, and EDA tool expertise or simply design resources. As such, contract services enable the design using the SIP product; the engineering work is peripheral to the SIP product itself.

Contract services usually require an initial fee paid upon execution of as SOW that describes the scope and nature of the services to be provided, followed by subsequent fees paid on a milestone or hourly basis. The SOW is usually attached to the SIP License Agreement.

There is no clear distinction between NRE and contract services in the SIP products industry. Occasionally some NRE development may include work peripheral to the SIP product and some contract services may include work that is central to the SIP product. Buyers should expect that contract services are essentially paying for access to resources and expertise to ensure successful use of the SIP products in the context of the EDA design flow, or in the context of the end product.

Design services are usually offered by EDA vendors, or design service houses often referred by foundries. Examples of contract services include EDA flow services such as, synthesis, analysis or place-and-route; custom design views such as timing and physical models; and IC integration services such as chip assembly, system verification and design for test or manufacturing.

7.8 Examples by SIP Product Type

An SIP product transaction usually combines several of the above business models and enabling components. Furthermore, the choice of business model may vary based on the type of SIP. Figures 7.7–7.11 provide examples by category, digital, memory, library and analog SIP, respectively.

[2] While we have mentioned revenue recognition issues in brief, any detailed discussion is well beyond the scope of this chapter.

Digital SIP: Digital SIP Products may be offered in several forms, soft, firm and hard. Figure 7.7 provides examples on per design, time-based and royalty models for the three forms.

Digital SIP	Per-Use	Time-Based	Royalty-Based
IP & Deliverables	Soft Silicon-proven DSP Netlist, Test Bench, etc.	Firm Silicon-proven PCI EDA Views + Test Bench	Protocol Processor EDA, GDSII, Firmware
Initial Fee – Scope	Fee for first IC in 0.18μ TSMC, UMC or CSM	Up-front fee for 3-years multiple use in UMC L130	Fee due on GDSII delivery Any IC in TSMC 0.13μ
Event Fees	Fee due on execution of SIP license agreement	Fee due on execution of SIP license agreement	Quarterly payments made after first shipment
(Re) Use Fees	% Reuse fee up to 3 ICs	Renewal upon expiration	Royalty per chip (% ASP) based on performance
Maintenance	15% covers bug fixes and minor spec updates	15% covers bug fixes and updates for rev 2.1 only	15% covers firmware and up to 2 process updates
Support	Incl: Email, tel, on-site AE	Incl: Email, tel, on-site AE, separate on-site R&D	Separate custom package
NRE	Functional modifications	Revisions beyond 2.1	Port to a new foundry
Contract Services	System level verification	Back-end implementation	Full chip verification
Source: FSA			

Figure 7.7 Examples of Digital SIP

The DSP core is a silicon-proven soft SIP product (netlist level). For example, it is licensed per design with the right to manufacture on TSMC, UMC and CSM 0.18-micron processes. The initial fee is due upon delivery of the EDA views (netlist, test bench, etc.) for the first chip. A discounted reuse fee applies for subsequent use for up to three chips. Maintenance is set at 15 percent of the license price and is not discounted; (it covers bug fixes, minor changes in specifications and basic support. Any functional modifications beyond the spec are treated as NRE. Verification expertise is offered on a contract service basis.

The Protocol Processor is a high-performance hard SIP product licensed on a per-chip basis with the right to manufacture in TSMC's 0.13-micron process technology. Customers may gain access to the front-end EDA views on a trial basis at no charge. The initial fee is due upon delivery of GDSII followed by a subsequent (optional) tapeout fee per design. Per-chip royalties are agreed up-front as a percent of ASP provided that certain performance criteria are met. Maintenance covers firmware updates but only two process changes, and support is priced separately. Any ports to a new foundry are treated as NRE and contract services include full chip verification.

The PCI core is a silicon-proven SIP product licensed for multiple uses with the right to manufacture in UMC L130 process. The initial fee is due upon delivery of EDA views and can be renewed in 3 years. Optional discounted fees may be structured on an annual basis. Maintenance covers only revision 2.1 and basic support. Any new revisions of the standard during the license term are treated as NRE. Back-end implementation is also offered as a contract service.

Memory SIP: Memory SIP is offered in many different ways, a single memory instance or memory configuration, a set of configurations to be used on a project or multiple uses from a memory compiler coupled with bit cells from a foundry or a provider. When memories use a foundry bit cell they are often restricted to use on that foundry.

Figure 7.8 provides examples on per-use (Instance), time-based and royalty models for different memory configurations.

Memory SIP	Per-Use (Instance)	Time-Based	Royalty-Based
IP & Deliverables	Dual-port SRAM 2K × 16 Col Mux 4 EDA Views, BitCell*	SRAM Compiler, EDA Views, and BitCell GDSII*	Custom High Speed SRAM EDA Views, BitCell GDSII
Initial Fee – Scope	Fee upon delivery Multi-uses in TSMC* 0.18μ	Up-front fee for multiple uses in CSM* 0.18μ	% Fee upon SOW Per design UMC L130
Event Fees		Annual TBL for compiler	% Fee upon tape-out
(Re) Use Fees	% of License fee to reuse on another foundry	Renewal upon expiration or for a new foundry port	% of chip cost based on acceptance criteria
Maintenance	Optional – 15% per annum	15% covers compiler, EDA, bitcell, process updates	12% covers bug fixes and up to 3 process updates
Support	Incl: Email, phone AE	Incl: up to three chips	Custom package
NRE	Characterizations and modifications if needed	Re-characterization	Custom development
Contract Services		Custom EDAviews	IC integration +testing

Foundry provided bitcell in GDSII
Source: FSA

Figure 7.8 Examples of Memory SIP

The dual-port SRAM is a silicon-proven hard SIP product, which is offered on a per-use model for unlimited usage of a unique configuration, in this case a 2Kx16 dual port (2rw), Column Mux 4 (determines aspect ratio). This can be duplicated multiple times on a single chip or across multiple chips. The SIP uses a foundry specific bit cell, so it is not portable. The license fee is paid upon delivery and a discounted reuse fee is charged for changing foundries. Maintenance is optional and costs 15 percent of the license fee per annum paid and accounted separately. Support covers e-mail, or applications support by phone. The SIP is delivered from a memory compiler, so there are no NRE charges unless there is a speed or power requirement that cannot be met with the standard part. Contract services are generally not required since the memory I/Os can be modified by the compiler.

The SRAM compiler and related bit cell is offered with a time-based model for multiple uses, and the right to manufacture in Chartered's 0.18-micron process. Since the buyer may not know its future needs in advance, the memory compiler allows reconfiguring the SIP for each design. The SIP is not portable since it uses a foundry specific bit cell (which may require a separate license from the foundry). The initial fee is due upon delivery of the SIP package (EDA views, GDSII, and compiler). There is also an EDA TBL component for the compiler that is paid annually. Maintenance covers compiler updates, EDA views and process changes.

The non-volatile RAM (NVRAM) is a custom developed component (NRE), which is licensed on a royalty basis based on agreed acceptance criteria in UMC's L130 process technology. The initial fee is due upon execution of the SOW, followed by a tapeout fee. Royalty is agreed to be a percentage of the cost per unit. Maintenance covers up to three process updates and support is separate. Any new configurations are treated as NRE, and any re-configuration for yield improvement as a contract service.

Figure 7.9 provides examples on per-use (project), royalty-based and access models for different memory configurations.

Memory SIP	Per-Use - (Project)	Royalty-Based	Access
IP & Deliverables	All SRAMs needed for a particular mask set	Self Testable, repairable, SRAM platform system	All memory compilers and complete bitcell portfolio
Initial Fee - Scope	License fee upon delivery for a specific mask set	Up front License Fee plus Royalties Per Unit	Up-front fee for 3 years in TSMC and CSM 0.18μ
Event Fees			% Fee per design due upon tape-out
(Re) Use Fees	% of License Fee to reuse on another mask set	% per unit sold, or per die based on savings (repair)	Renewal upon expiration
Maintenance	Optional 15% per annum	15% covers bug fixes and process updates	Included
Support	Incl: Email, tel, on-site AE	Incl: Email, tel, on-site AE	Incl: Email, tel, on-site AE
NRE	Characterizations and modifica tions if needed	Characterizations and modifications if needed	Re-characterization
Contract Services		Test + repair integration yield enhancement	Custom fit sizes

Foundry provided bit cell (implicit time base)
Source: FSA

Figure 7.9 Examples of Memory SIP

The pre-determined number of unique SRAM configurations is licensed on a per-use basis with the scope of use to be a particular project. The project is defined as a mask set for a particular foundry, process node and variant (TSMC 0.18G). Since the number of configurations may vary, the project may be priced as a small, medium or large project based on the number of configurations needed. Configurations can be used multiple times on the mask set. The initial license fee is paid upon delivery, followed by a discounted reuse fee for another mask set (subsequent to initial use). Maintenance is optional, and if selected is 15 percent per annum and includes standard support. NRE charges apply for modifications and/or characterizations.

The self-testable and repairable memory system is a high-density, high-yield platform offered in pre-determined unique SRAM configurations. The different configurations include redundancy, a self-test and a repair processor in soft SIP form and a fuse box in hard SIP form. The memory system is intended for use on a particular project which is defined as a mask set for a particular foundry, process node and variant (TSMC 0.18 μm), with multiple instantiations on the mask set. The initial license fee is paid upon SIP delivery. Royalties are structured per unit sold, or as a percentage of the savings due to yield improvement upon repair. Customers may want assistance integrating the memory system into the rest of the chip, or with the actual test and repair operation. A yield enhancement service including failure analysis may be provided to improve yield.

The complete memory compiler and bit cell portfolio package is licensed for multiple uses with any 0.18-micron process in TSMC and CSM for a period of 3 years. The up-front access fee provides flexibility to create any SIP configuration. Subsequent discounted fees are paid for each SIP configuration on per design basis upon tapeout.

Library SIP: Foundation library or specialty library SIP is usually offered as a package including standard cells, I/O pads and memory bit cells from the provider, the foundry or both. Figure 7.10 provides examples on per-use time-based, third-party royalty-based and access (including per-use) models for the different forms.

The high-performance library package is licensed on a per-use basis for UMC L130 process technology. An initial fee is paid by the end-user upon SIP delivery followed by a discounted fee upon tapeout. Additional reuse fees apply for up to three more chips. Maintenance covers only two process updates and support is priced separate based on customer needs. Any customization for I/O cells is structured as NRE.

The full standard cell, I/O and bit cell package for TSMC 0.18-micron process is offered on the provider's Web site through a click-wrap license (i.e. download) at no charge to the customer. The SIP is not portable (contains foundry specific bit cell and I/O) and the license includes provisions for the right to audit and a separate license required from TSMC. Royalties are paid to the provider by TSMC on a per wafer basis. The provider has a separate

Library SIP	Per-Use	Royalty (3rd Party)	Access (+Per-use)
IP & Deliverables	High speed Std Cell Lib and I/O package	Std Cell Lib, I/O*, Bitcell*, & RAM compiler package	Complete Cell, I/O, SRAM Compiler, and bitcell* pack
Initial Fee – Scope	First chip in L130	Free for TSMC 0.18μ (Paid by foundry)	Annual access fee for use in UMC L180, L130
Event Fees	Tape-out fee (paid by end-user)		Per design Tape-out fee (Paid by Fabless ASIC Co)
(Re) Use Fees	% of Initial fee for each new chip up to 3 uses	Royalty per wafer (Paid by foundry; Audit right)	
Maintenance	12% covers up to 2 process updates	Free; Covers up to 4 process updates	Included; Covers process updates beyond time
Support	Paid separately	Applications support package paid separately	Custom premium package
NRE Charges	Custom I/O cells	New custom cells	I/O modifications
Contract Services		Special characterization	

Foundry specific Foundation SIP
Source: FSA

Figure 7.10 Examples of Library SIP

agreement with TSMC for distribution and support of their SIP and offers separately priced application support packages.

The full library package for UMC L130 and L180 processes is licensed in a combination of access and per-use models to a fabless ASIC company. The access fee is paid annually, with subsequent per design fees paid upon tapeout by the fabless ASIC company who has rights to sublicense the SIP to its customers. Maintenance is included in the price, custom support is paid separately and I/O modifications are treated as NRE.

Analog SIP: Analog SIP is always offered in hard form (GDSII) for a specific process technology and is highly dependent on process technology updates and/or change in SPICE models. Figure 7.11 provides examples on per-use, time-based and royalty models.

The PLL SIP package is licensed per design as an SIP configuration (one or multiple PLL instances may be used in a chip) for Chartered's 0.18-micron process. The initial fee for the first chip is paid upon delivering GDSII, followed by a subsequent discounted fee upon tapeout. Each new chip for the same foundry and process technology follows the same model. Maintenance covers only up to two process updates for minor layout modifications and characterizations, and a basic support package is included. Any port to a new foundry or process is treated as NRE and results in a new SIP license.

Analog SIP	Per-Use	Time-Based	Royalty-Based
IP & Deliverables*	50–250 MHz PLLs (Multiple SIPs per config)	Dual 6-bit and 10-bit low voltage DACs	Custom 10-bit 400MHz DAC optimized for video
Initial Fee – Scope	First chip for CSM 0.18 μ	Perpetual for multiple uses in UMC L180	% upon execution any chip in UMC L130
Event Fees	% upon Tape-out		% upon SIP acceptance % upon Tape-out
(Re) Use Fees	New chip in CSM 0.18 μ (Alternate SIP config)	Renewal upon expiration	Fixed fee per chip within 30 days of quarter sold
Maintenance	Included for first year 2 process changes max	15% paid annually covers only bug fixes	18% covers process changes beyond term
Support	Incl: Email, tel, on-site AE	Incl: Email, tel, on-site AE	Separate for on-site eng
NRE Charges	Port to new foundry (New SIP License)	Re-characterization	% upon GDSII delivery
Contract Services		Custom size modifications	Platform integration

*All deliverables are in GDSII
Source: FSA

Figure 7.11 Examples of Analog SIP

The low-voltage DACs are offered in a perpetual time-based model and multiple uses for any chip in UMC L130 process technology (implicit time base). However, the license expires if the customer wants to use the SIP on another foundry or process node. Maintenance covers only bug fixes and basic support. Any characterizations needed due to process changes are treated as NRE (enable the SIP itself), and any custom size modifications (enable the IC design) are treated as a contact service.

The 10-bit video DAC is a specialty SIP that is enabled through an NRE arrangement, and licensed on a combination of per-use and royalty models. A portion of the NRE fee is paid upon execution of the agreement with subsequent portions due on milestones (i.e. GDSII delivery, or acceptance criteria met). Upon acceptance, the SIP is licensed on a per design tapeout and per IC unit combination. Royalty is structured as flat fee per unit (priced as a percentage of the cost of an equivalent pure analog IC minus the packaging costs). Maintenance is paid separately at 18 percent of the license fee and covers basic support and process changes beyond the term of the license. A contract service may be a platform integration of multiple SIPs into one to reduce die size. This model may be used for custom SIP that may be expensive to develop and maintain, and may require special acceptance criteria due to sensitivity in process changes.

These examples should provide sufficient insight on how the business models, enabling components and combinations of the above operate in the current SIP products industry.

The SIP products business is unique and depends on a variety of factors making it difficult or impossible to standardize on a single business model.

Furthermore, the wide variety of uses coupled with the different forms and types of SIP makes it difficult to understand how to extract SIP value and negotiate an appropriate licensing agreement. By establishing a common terminology on business models, objective comparisons become easier and thus the negotiation and licensing process are simplified.

7.9 Licensing SIP Products

SIP License Agreements may require a significant effort depending upon the business objectives of the parties, the intended use of the SIP Product, the nature of the SIP product and the risks associated with using the SIP product in the end application.

Buyers sourcing SIP products from multiple providers must cope with a variety of business and licensing models among providers. Each SIP License Agreement is unique and therefore the process can be very time consuming, often taking as much, or more, time than the technical evaluation and integration. The process can be simplified by establishing a common taxonomy for SIP licensing and some understanding on what to expect based on different transaction scenarios.

The key to a successful licensing negotiation is to know the issues and the likely positions of the other parties involved in the transaction. It is essential that these positions be fully discussed early in the transaction so that no surprises emerge late in the negotiations.

7.9.1 SIP Licensing Provisions

Figure 7.12 provides a high-level description of the common provisions found in an SIP License Agreement, a short description and the consensus on the level of effort likely to be involved in negotiating these provisions.

An SIP License Agreement contains provisions that are generally included in most agreements (marked as "R" for required) and others that may be included (marked as "O" for optional). Some of the optional provisions may be required if the scope of the SIP License Agreement extends to additional deliverables or services. For example, if some NRE work were involved, then the buyer would probably require an acceptance provision and both parties would probably negotiate some special conditions.

Provisions that tend not to vary much are generally placed in the main body of the SIP License Agreement. Others that tend to be more idiosyncratic such as payment structure and special conditions tend to be found in the appendix. The SIP License Agreement is negotiated and

Contract Provision	Description	Where	Reg/Opt - Effort
Definitions	Type of agreement; Parties involved; SIP specifics; License grant rights	MLA	R
Scope of License	Where and how the SIP can be used *Use, Copy, Modify, Distribute, Sub-license*	MLA App	R
Restrictions-Rights	"Forbidden uses" of the SIP Product; Disclaimer on third party licenses	MLA	O
Deliverables	Items are delivered; Media of delivery; Terms an conditions	MLA App	R
Acceptance	Conditions under which a non-silicon-proven SIP product is accepted	App	O
Payments	Fee structure; Payment terms; Method; Currency & taxes; Rights/Terms of audit	App	R
Special Conditions	Requests related to SOW; SIP Mobility; SIP Performance; Sales & Marketing	App	O
Warranty	How long is the coverage; What to do if SIP does not work; What voids warranty	MLA	R
Indemnification	IPR covered; Remedies and obligations if a third party alleges SIP infringement	MLA	O
Limitation of Liability	What happens if use of an SIP causes harm to the Buyer; Formula for limitation	MLA	O
Maintenance	What is covered or not; Which process-tool updates; What if the spec changes	MLA	R
Support	What is included or not; Other contract for support based on Buyer's needs	Other	O
Term and Termination	How long the agreement lasts; What are the conditions for early termination	MLA	R
Confidentiality	What is confidential; How it's being treated; How to keep it that way	MLA	O
General Provisions	Governing law; Assignment; Export requirements; Integration/modification	MLA	R

Sources: VCX model agreement; Keith Witek - AMD; Carl Hoxeng - Virage; SIP provider agreements
Legend: R: Required; O: Optional; MLA: Usually in Master License Agreement; App: Usually in Appendix
Source: FSA.

Figure 7.12 Typical SIP License Provisions

signed once upon initial purchase. The appendices are typically negotiated and signed with each purchase. Many of the provisions of a typical SIP License Agreement are straightforward and do not tend to require a great deal of negotiation. There are some provisions that take more effort than others do; the longer bars in the effort column indicate these.

7.9.2 Definitions

A typical SIP License Agreement starts by identifying what type of agreement is being formed, the date when the agreement begins – sometimes called the effective date, the names of the parties and their location. This section may also include definitions of certain specifics related to the SIP product, such as SIP form, EDA views and target technology. It may also define any third parties who may be involved in the SIP License Agreement such as foundries, design houses and others.

7.9.3 Scope of License

The Scope of License provision states what the buyer can do with the SIP. Normally the Scope of License provision will specify how, if at all, the buyer can, use, copy, modify, distribute and sublicense the SIP.

- *Use*: This provision specifies how the buyer may use the SIP product and how this is related to the provider's business model. This provision may also specify whether the buyer can use third parties such as design houses in furtherance of the rights granted to the buyer; in other words, it may specify whether or not the buyer may engage a third party to design using the SIP product. Normally if such third parties are to have access to the SIP product, this provision will spell out the restrictions that the buyer must impose on such third parties before they have access to the SIP product. In some cases, there may be an Addendum that lists the third parties, which may have access to the SIP.

- *Copy*: This provision specifies how many copies of the SIP product the buyer can make and the security restrictions on these copies. This provision is used to limit the provider's risk of unauthorized use or "leakage" of the SIP product. Normally, these are copies other than those made as part of the manufacturing process; it may be helpful to think of the copies restricted under this clause as copies of the SIP product's design files and documentation.

- *Modify*: This provision specifies what (if any) modifications the buyer is allowed to make to the SIP product. For example, if the SIP product is licensed in soft form, then the modification provision might prohibit all modifications except buffer changes and cell resizing. If the SIP product is licensed in hard form, then the provision might prohibit all modifications except changes in certain layers of the GDSII file, such as I/O pins.

- *Distribute*: This provision specifies how the buyer may distribute the SIP product. In most cases, this is a simple right to distribute a device or devices containing an instantiation of the SIP product, but it may be more complex. For example, a provider may license a library package to a fabless ASIC vendor and grant certain distribution rights that enable the fabless ASIC vendor to provide the EDA views and SIP product support to their customers.

- *Sublicense*: This provision specifies whether and how the buyer might be allowed to pass on to a third party all or a portion of the rights granted to them. This type of provision is common in original equipment manufacturer (OEM), reseller or distribution agreements, but is generally not that common in agreements where the buyer is simply given the right to design a device using the SIP product.

7.9.4 Modification Rights and Ownership

Most SIP License Agreements will state whether the buyer has the right to modify the SIP product. If so, then there is usually a provision in the agreement that specifies who owns the modifications. Normally this takes one of three forms: the buyer owns the modifications outright with some form of license back to the provider, the provider owns the modifications or there is some sort of shared ownership.

As noted previously, there may be some cases where the modifications – normally as part of an NRE contract – result in the creation of a new SIP product. There are many ways to allocate ownership in such situations. Some agreements approach ownership of the IPR in the new SIP product from an equitable or "sweat of the brow" theory; that is, the party who has done most of the modification is entitled to the lion's share of the IP rights in the resulting SIP product. Others may allocate SIP product ownership based on the individual core competencies of the two parties. In any of the above cases, there may be sublicense rights, particularly if newly formed SIP product contained SIP products from both parties.

7.9.5 Restrictions on Scope

Normally, following the scope of license provision in an SIP License Agreement, there is a provision defining the "forbidden" uses of the SIP product. For example, the restrictions on scope provision may contain a clause stating that certain layers of an SIP product in hard form used for device tracking purposes may not be changed. Other restrictions may include a prohibition on reverse engineering, decompiling of computer code, black box probing of the SIP product or disassembly.

In addition, common in this provision is a disclaimer of any implied licenses. Normally, this provision will state that the rights given to the buyer in the SIP License Agreement are all the rights the buyer receives and that any implied licenses[3] are excluded to the extent possible under law[4].

---

[3] In the United States and Europe, the law may imply certain statutory warranties dealing with, for example, title, non-infringement or merchantability.

[4] In some cases, the law of a particular jurisdiction may not allow certain implied warranties to be disclaimed.

Another issue that may also be dealt with in this clause is any requirement for licenses from third parties. Sometimes the use of an SIP product may require a separate license from a third party that providers may or may not be aware of. For example, the I2C serial interface: a buyer who needs an I2C SIP product may license the SIP product from a number of sources. However, to use the SIP product the buyer must also obtain a separate license. Some providers will state in the License Agreement that use of their SIP product requires the buyer to take a license from a named third party. Other providers may state that they are aware of the need to take a license from a third party but also caution that the buyer should find out if any other licenses are needed. Still other providers disclaim all responsibility and put the obligation on the buyer to find out if any third-party license is needed to use the SIP product.

7.9.6 Deliverables and Acceptance

This provision generally defines the items to be delivered that constitute the SIP product such as design files, EDA views and models. It may also state what acceptance testing criteria exist, if any. This provision normally specifies not only the items to be delivered but also at what point the buyer assumes the risk of loss for any such items and whether the items are to be delivered on a CD, other magnetic media or via electronic download. With respect to acceptance criteria, this provision tends not to be very common where the SIP product is silicon-proven. In such cases, acceptance normally occurs as soon as the items to be delivered are actually delivered. Providers generally favor this approach since it allows them to recognize most of the revenue when the SIP product has been delivered to the buyer.

If the SIP product is not silicon-proven, then there may be some acceptance criteria and procedure for the parties to determine whether the SIP product delivered functions according to the specification. For acceptance, there is normally a defined time period within which the buyer must indicate his acceptance or rejection of the SIP product; the criteria and time period may vary by SIP product type and by individual customer needs. For example, buyers might request special acceptance criteria to reduce their risk in using an unknown SIP product. However, as previously mentioned, providers tend not to favor special acceptance criteria as they may affect revenue recognition.

7.9.7 Payments

In all SIP License Agreements there is a provision stating how and when payments are to be made to the provider. Some elements of this provision may be put in the appendix to the agreement. Normally payment provisions define the structure of the fees paid and the currency in which payment is to be made. Most agreements will specify when payment is to be made and whether it is to be made by check, wire transfer or other means. For agreements that involve parties in different countries and payments taking place over an extended period of time, there may be some provision for any risk associated with fluctuations in exchange rates. Finally, if the buyer is in a country that strictly controls foreign exchange, the agreement should also state who is responsible for obtaining the necessary permissions or permits.

Payment provisions will also usually specify who is responsible for taxes associated with the transaction. While this may not be much of an issue in the United States and Europe there are certain countries where it is important to be aware of certain peculiarities in the local tax code[5].

If the SIP product is offered on a royalty basis, then the payment provision will also normally contain specific clauses relating to how royalties are to be calculated, how and when payments are to be made and how often the provider may audit the buyer to determine the correct payment of royalties.

It is worth noting that audit provisions are becoming increasingly common since providers need to track shipments of products using the licensed SIP products, particularly in per-use and/or royalty-based business models. Recent changes in accounting practices require public companies to keep records on royalties that may require audits.

7.9.8 Special Conditions

Special conditions usually appear in the appendix since they are specific to the needs of the parties. It is hard to generalize about what these special conditions may consist of, so the following list of examples should provide an idea:

- *SOW*: There may be an SOW for additional work that is to be performed as part of the SIP License Agreement; this may include milestones, an approval process for intermediate/iterative deliverables, state and regular progress updates.
- *Special SIP Requirements*: These provisions may impose special performance criteria for the SIP product as integrated into the end device, such as its support of certain corner cases or minimum yields.
- *Sales and Marketing*: These provisions may be common where the provider wants the buyer to become a reference customer, or where the provider may want the buyer's assistance in marketing or public relations.
- *SIP Product Mobility Requirements*: To ensure an alternate source of silicon, the buyer may require the provider's assistance in moving the SIP product to another foundry process if there is limited foundry capacity when the device goes to production.

[5] Readers may be familiar with the Withholding Tax issue that applies to buyers located in Japan and providers located elsewhere.

SIP product mobility provisions are becoming more common in SIP License Agreements. As buyers become more aware of the limitations inherent in using non-portable SIP products, they are increasingly requesting assurances from the provider on alternative sources of supply.

7.9.9 Warranty

Warranty provisions vary widely among providers, principally because there are, as of yet, no objective standards for SIP product performance in the SIP products industry. Typically, the warranty section of an SIP License Agreement will deal with the following issues:

- *Basis of the Warranty*: What forms the basis of the warranty? This is typically the provider's policy on the functional and operating performance for the SIP product and is normally contained in some objectively referenced document such as a data sheet.
- *Duration of the Warranty*: It is difficult to generalize what an appropriate duration is for an SIP product's warranty. Providers tend to keep warranty obligations as simple as possible to simplify the revenue recognition process. At the same time, buyers tend to seek longer warranty periods to ensure efficient deployment of the SIP product.
- *Notice of Non-Compliance*: Normally the warranty provision will describe how the buyer must notify the provider of any defects and the time period within which the buyer must notify the provider of any non-compliance.
- *Provider's Obligations on Non-Compliance*: If the SIP product does not meet the warranty criteria, the warranty provision describes how the provider will remedy such non-compliance. Usually this means the provider fixing or replacing the SIP product within a specified time period. While some SIP License Agreements may also mandate that the provider refund the license fees paid if these remedies fail, this provision is not common.
- *Situations That May Void Warranty*: The warranty provision of an SIP License Agreement will normally contain a provision that states that if the buyer will void the warranty if he engages in certain conduct, such as unauthorized modification. Normally this is justified on cost grounds: providers simply cannot support buyer modifications where the provider does not necessarily know what modifications were made.
- *International Implications*: Depending on the jurisdiction in which the buyer is located there may be mandatory warranty requirements that cannot be avoided by contract. Providers should be well aware of such requirements before licensing SIP products in such jurisdictions as they can have a significant effect on pricing.

7.9.10 Indemnification

In the context of an SIP License Agreement, indemnification refers to the obligation of the provider to make the buyer whole if the buyer is not able to use the SIP product because the SIP product infringes (or is alleged to infringe) some third party's IP rights. Typically, the indemnity section of a License Agreement deals with the following issues:

- *IPR Covered*: The provision normally specifies which IP rights are included within the indemnity. In many SIP License Agreements, the coverage is typically for copyrights, trade secrets and patents, but there are some agreements that specifically exclude patents.

- *Applicable Geography*: The provision normally specifies which country's patents, copyrights, trade secrets and the like come within the indemnity. The question is usually one for negotiation and normally relates to the location of the parties and the likely market for devices containing the SIP product.

- *Provider's Obligations in the Event of Infringement*: This provision normally specifies what the provider is obligated to do if there is actual or alleged infringement. It will typically state that the provider must defend or settle the action and provide some form of remedy to allow the buyer uninterrupted use of the SIP. In general, this will involve the provider taking a license from the third party alleging the infringement, or the provider redesigning the SIP product or providing the buyer with a functionally equivalent but non-infringing SIP product.

- *Buyer's Obligations in the Event of Infringement*: This provision normally specifies what the buyer is obligated to do if there is actual or alleged infringement. Normally the buyer must cooperate with the provider in any defense or settlement of the action and not do anything to prejudice the provider's defense or settlement of the claims.

- *Conduct That Renders the Indemnity Null and Void*: This provision normally specifies what conduct by the buyer, such as unauthorized modifications or uses, may void the indemnity.

7.9.11 Limitation of Liability

Another critical issue that is sometimes intimately linked to the indemnity provision is the limitation of liability. This provision typically states the maximum liability of the provider if the business transaction causes specified harm to the buyer. Typically, the limitation of liability section of an SIP License Agreement deals with the following issues:

- *Formula for Calculating the Limitation*: A typical formula is license fees paid over the life of the license. However, this formula may be negotiated between the parties to contain one or more of the following variations:
  - A multiple of license fees.
  - License fees paid over a specific period of time, such as the 12 months preceding the event, this is especially common in licenses of long duration.
  - A "floor" below which liability will not go. This is common in multi-year licenses where payments taper off and using a formula of the preceding 12 months would produce a limitation of liability figure of zero.
- *Exclusions from the Limitation*: The parties may wish to exclude certain conduct or certain provisions of the agreement from the limitation of liability. In many SIP License Agreements, buyers will seek to exclude the indemnity provision from the limitation of liability; in other words, the buyers want the provider's liability for third-party infringement to be unlimited.
- *Reverse Indemnities*: Certain conduct by the buyer might result in liability to a third party for the provider; in some cases, the provider will want the buyer to indemnify them from any damages resulting from such conduct.

In limitation of liability discussions, there is often a significant gap between the limitation providers are willing to provide and the limitation buyers wish to have. Providers typically want to limit their liability to the amounts received under the license, while buyers want to have a limitation of liability that bears a greater relation to the risk they are taking by using the SIP product. The result is often a prolonged and contentious negotiation. Often the limitation of liability question is the last issue to close in a negotiation and leaves both parties feeling dissatisfied.

Even the most established providers are not able to bear the potential financial burden of millions of dollars in damages for an SIP product license from which they have only received tens of thousands of dollars. This gap between the limitation providers will offer and the limitation buyers expect is called the indemnity gap. While this is certainly not a new problem, there are solutions that are beginning to bridge this gap.

7.9.12 Maintenance and Support

Maintenance, which may also include basic support, is usually in an appendix. The parties should ensure that this section clearly spells out what is covered and what is not covered.

In addition, providers may include some special provisions for accounting purposes because maintenance is not discounted and the term of maintenance that may be different from the term of the SIP license.

Support, apart from predefined packages, is usually structured in a separate contract that may be referenced in the SIP License Agreement. No matter where it may be located, buyers should ensure that it clearly spells out what is covered and what is not covered.

7.9.13 Term and Termination

This provision generally states how long the agreement lasts, when it ends and what conditions can bring it to a premature end. Another term that may be found in the termination provision deals with the buyer's rights after the agreement has terminated. For example, what may the buyer do with devices that have been manufactured using the SIP product but have not been sold? Termination provisions will typically also include any requirements for ongoing support and the like.

7.9.14 Confidentiality

A confidentiality provision is usually found in most SIP License Agreements. It is generally designed to protect trade secrets[6] and will typically define what information the parties deem confidential, what standard of conduct must be used for to maintain the confidentiality of the information and how long the obligations of confidentiality will endure. Typically, these obligations can be expected to last for between 3 and 10 years following the expiration or termination of the license.

7.9.15 General Provisions

General provisions in SIP License Agreements usually include governing law and jurisdiction, assignment, export requirements and the rules and conditions under which the agreement can be modified.

A choice of governing law can sometimes be overlooked in SIP License Agreements. It is advisable to make sure that the governing law is always stated clearly to avoid unexpected issues and difficulties that may arise, particularly in foreign countries where the environment is unfamiliar.

Assignment provisions are generally designed to ensure that the provider is aware of the buyer's desire to assign its rights and obligations under the SIP License Agreement and has

[6] Unlike patents and copyrights, trade secrets (also known as know-how in certain jurisdictions) are of theoretically infinite duration. However, trade secret protection is lost if the information is revealed to a party without an obligation on that party to keep such information confidential.

the right to veto any such assignment if the assignment would be prejudicial to the provider's business. Normally these provisions do not receive much scrutiny but they can be important especially where the provider wishes to avoid the license being assigned to a competitor.

Export requirements typically state that some or all of the technology in the SIP product may be subject to requirements placed on sensitive technology by the US Department of Commerce's Export Administration Regulations (EAR) and the US Department of State/Department of Defense's International Traffic in Armaments Regulations (ITAR). They generally place a requirement on the buyer to comply with the requirements of these regulations when using, licensing or otherwise transacting in the SIP product.

Finally, there is usually a so-called "Merger" or "Entire Agreement" clause. This provision states that the SIP License Agreement is the entire agreement between the buyer and the provider. It will typically also state that all oral understandings – promises, sales commitments, marketing "puffery" and the like – not documented in the SIP License Agreement are null and void and that any other written documents such as purchase orders and the like shall have no effect on the License Agreement. It will also typically contain a provision that specifies how if at all the License Agreement may be amended.

7.10 Provider and Buyer Perspectives

The scope of license, warranty, indemnity and limitation of liability provisions usually consume the majority of the effort in negotiating an SIP License Agreement. As noted above, buyers and providers tend to have divergent views on how these provisions should be approached. Buyers may seek a more consistent policy on the above areas as they cope with multiple providers and SIP products instantiated in a single design. Figures 7.13 and 7.14 outline some of the key areas where the buyer and provider perspectives often diverge. While it might be easy to see how an SIP user might ask for or require some generous licensing, deliverable, support or warranty provisions from an SIP provider, it is important to understand that some of the terms may have severe implications on revenue for the provider.

- *License Grant*: Providers usually seek a narrow scope of license to maximize revenues, while buyers try to broaden the provision to maximize flexibility. Providers may view a new scope of use as a new product or license fee, while buyers may feel that the additional cost may not be justified. For example, some providers view that a new port for an SIP product is a new use and thereby treated as new SIP product that justifies an additional fee if not a new SIP License Agreement.
- *Restrictions*: Providers usually seek to impose as many restrictions as possible on the scope of use to minimize cost and risk, whereas buyers seek as few restrictions as possible to maximize use, reuse and distribution of the SIP product. As part of their broad restrictions, providers may seek to limit the countries in which the buyer may

use or distribute the SIP product, or may seek to limit the number of copies of the SIP product the buyer may make, and place restrictions on modifications and assignment, distribution and sublicensing of the SIP product. Buyers usually seek broader rights to modify, reproduce and distribute the SIP product.

	Provider Perspective	Buyer Perspective
License Grant	Narrow scope of use	Broad + flexible scope
Restrictions	Limit geographies, copies, etc.	Few or no restrictions
Deliverables	Set of items; Accept on delivery narrow maintenance coverage	More items; Accept on criteria; Broad maintenance coverage
Warranty	As is; Applies to SIP only Above and beyond is NRE	No limitation; applies to SIP and IC
Remedy – Cure	Fix, replace – weeks	Fix, replace, refund – days
Period – Parties	30–90 days Licensee only	3–5 years licensee + related parties
Source: FSA		

Figure 7.13 Perspectives on License Grant and Warranty

- *SIP Deliverables*: Providers may tend to view the deliverables as limited to the set of items for the SIP product (current process technology, version and/or specific EDA tools). As previously mentioned, they tend to seek acceptance of SIP product upon delivery to the buyer. Buyers may seek a broader set of deliverables and special acceptance conditions beyond delivery of the items. It must be noted that any acceptance criteria imposed by the SIP user will delay the ability of the provider to recognize the revenue from the licensing transaction.
- *Maintenance and Support*: Providers usually offer a small number of updates and separate support from maintenance because the scope of use varies from customer to customer. Buyers, on the other hand, seek much broader maintenance coverage (full conformance to spec, collateral, future versions, updates, upgrades and the like) to reduce cost and risk. As noted earlier, the maintenance and support agreement are generally separate from the licensing agreement. Again, this is because the maintenance obligations of the SIP provider are treated differently from an accounting standpoint. Simply including the maintenance or support obligations in the license agreement will require the provider to "back out" some portion of the contract value and delay it for the period of the support obligation.
- *Warranty*: Providers seek to limit warranty to the performance of SIP product itself. They usually ask for longer lead times to cure any problems and may try to limit the warranty period to a few months. They may also seek to narrow the warranty coverage

to the buyer only. At the same time, buyers seek a longer warranty. They normally seek to minimize the risk that the end product will be delayed if the provider's cure period is long. Buyers may also try to reduce risks associated with end product performance (if there are issues with the SIP product used in the context of the IC); and they try to increase coverage with respect to end product life cycle (if the warranty period is short). Both parties need to carefully weight the pros and cons on these complex issues and appreciate the fact that the SIP product industry has not yet matured enough to support ISO compliance standards. As with maintenance and support, if a warranty provision extends beyond a typical 90-day period, the revenue from a licensing agreement will be reduced to account for possible problems in the future. Thus, the tendency of the SIP providers to focus on a maintenance agreement to address problems that arise after the initial license agreement is signed.

- *Indemnity*: From the provider's perspective, a broad indemnity provision can be problematic. Although a provider can be fairly confident that the SIP product does not infringe known patents, it is virtually impossible to be certain because much of the captive SIP in the semiconductor industry has not been commercialized. Furthermore, it is very difficult to be certain if a third-party contractor who contributed to the SIP product does not infringe the IP rights of another party. As a result, providers seek to narrow the scope of indemnification, geographies and remedies, while buyers try to broaden the scope to minimize risk. Again, there is not an immediate solution in the SIP products industry. Good faith collaboration between the parties in the event of an infringement combined with some creative insurance solutions may be the most productive remedy at the moment.

	Provider Perspective	Buyer Perspective
Indemnity	Copyrights + trade secrets indemnify only	No limitation on any SIP defend, indemnify, hold harmless
Geography	Limited the US, EC, and Asia	Worldwide without restriction
Remedies	License to use, design around, refund, termination	No restrictions, some remedies
Liability	Based on transaction value	Based on business risk
Damages	Limited to license price, or multiple with a cap	No caps or restrictions
Exclusions	Improper-use; modifications	Flexibility trade-offs

Source: FSA

Figure 7.14 Perspectives on Indemnity and Limitation of Liability

- *Limitation of Liability*: Providers and buyers often face a major negotiation gap with respect to potential financial losses resulting from the use of the SIP product, which may be functional failures and/or third-party IP infringement. Providers view that their liability should not exceed the value of the transaction whereas buyers view that the value should be based on the business risk. The formula for damages is difficult to calculate and varies based on SIP product type, its level of maturity, and provider's risk management profile. Typically, providers will limit damages to license fees received – or at most, two to three times the license fees received – and tend to spell out any exclusion that void this provision, which, in turn, may limit flexibility for buyers. In some cases, providers may be willing to take more risk if their SIP product is silicon-proven across several customers, but this is not the industry norm.

7.11 The Evolution of the IP Industry

The time and cost benefits of using SIP products is real and can be substantial. However, technical standards, product quality, consistency in business practices and licensing agreements still need to improve for the industry to enjoy a true plug-and-play commercial SIP product trade throughout the supply chain.

From a supply chain perspective, there are several business and technical dimensions, which influence the SIP product industry's ability to standardize on business models and licensing provisions. To reduce these barriers providers, EDA vendors, foundries, IC vendors and their customers must work closer with one another and provide valuable input to industry consortia and standards committees. Only then, will the SIP products industry grow and the entire supply chain will benefit.

The use and reuse of existing commercial IP is enabling improvements in time-to-market and cost reduction. The number of IP providers in the electronics industry is growing rapidly. The number of IP provider companies has become as important and prolific as the emergence of another industry-driving force, the fabless semiconductor companies.

IP – everything from complex cores and libraries to specialized analog blocks – is an integral part of the fabless semiconductor industry today. The transition of commercial IP from being a novelty to the norm has created significant shifts as the industry matures. Product and business models are converging, and the industry is consolidating and stratifying into a few major players and many smaller "niche" players. Maturation and consolidation mean that the IP industry's technology is accelerating beyond the capabilities of all but the largest semiconductor companies.

The major players in the IP industry now have the size and technology resources to perform advanced research and thus become technology drivers in the semiconductor industry. This section examines how and why these trends are occurring, how this will result in a broader, less redundant IP portfolio for the industry and the impact on the fabless design community as well as the pure-play foundry market.

7.11.1 The Virtuous Spiral

IP – and especially commercial libraries – have long been a part of the fabless industry. Many would argue that commercial libraries helped enable the fabless model. The value proposition was simple: instead of using limited cash flow in hiring people and buying EDA tools to build libraries and IP, a fabless company could, instead, purchase IP from a commercial supplier.

Theoretically, IP suppliers' leverage from selling the same IP multiple times should make their products far less expensive than un-leveraged single-use components. The benefits of leverage can be quite large for broadly used components such as libraries, bus interface cores and other IP. Theory and practice, however, are often far apart. Until recently, commercial IP was not dramatically less expensive than blocks built for a single use. Why? The primary reason for cost differences was that apples were being compared to oranges. IP suppliers typically pour more resources into a reusable block than someone building a single-use block. The product is more complete, including far more extensive validation and testing, additional EDA tool models and other additional features. Since quality is by far the most important, many IP companies – including Artisan (since acquired by ARM) – invest enormous amounts in rigorous product development methodologies, integration and accuracy within the EDA tool flows and other tests. For all these reasons, commercial IP is commonly more robust and feature-rich than single-use cousins. Added effort creates high-value IP and this value is leveraged across all users.

The other cost addition was transaction costs. Unlike IP development, however, these costs do not add much value, nor are they able to be leveraged well. When IP is purchased directly, transaction costs can be a significant portion of the total IP cost. These costs include application engineering, marketing and sales activities whose goal is to entice designers to select a particular product – and then pay for it. Evaluation goes far beyond "fitness for use" attributes of product quality, high-performance and ease-of-use, and instead focuses on cost/feature comparisons among competing products.

The IP industry has focused on reducing transaction costs by leveraging economies of scale, company reputation and "try-before-you-buy" relationships. The extreme example of reducing transaction costs is the "Free IP" model, about which much has already been written. The "Free IP" model eliminates most transaction costs between the IP supplier and the end-user: end-users can use IP for no up-front charge and typically no usage or tapeout charge. The foundry compensates the IP supplier when a chip goes into volume production.

The transaction cost for each additional user is virtually non-existent, yet return is directly proportionate to usage. Whether the IP is free or includes some up-front payment, the emerging IP business models align the goals of the customers and the vendors by shifting the lion's share of the economic benefit to the end of the cycle, triggered when and if the design team's chip goes into production. Everyone is successful together, at the same time, or not at all.

As IP proliferation increases, IP prices drop, causing a curious switch in the IP adoption model. In the traditional model, when a design company purchased IP, the task was for the IP vendor to prove sufficient superiority to justify the expenditure. Now, the task is reversed: the company must justify the expense of internal development versus using low-cost or no-cost commercial IP. Quality, performance and reliability are all still vitally important, but a widely used commercial product will quickly establish itself – or not.

The lower the price of the IP, the faster the adoption. For example, Artisan Components' free library program distributed over 300 libraries to users worldwide, which would have taken a decade in the old up-front purchase model. In a high-proliferation environment, where product revenue funds IP development rather than channel costs, a "virtuous spiral" is created: increasing product usage by customers leads to greater return on the IP, leading to greater investment by the IP supplier, creating even higher differentiation, leading to even greater usage and so on.

Market share becomes the key. All of this can shift very quickly – relative market share can change dramatically in just a few months when radical improvements to products or business models become available. Since the market will support only a few competing offerings in an IP product category, a handful of IP companies will emerge as large category leaders. These companies will make large investments in all aspects of the product – technology, EDA tool infrastructure, ease-of-use and third-party relationships – which will be out of the reach of smaller companies. These large IP companies will provide the "mass-consumption" IP – processor cores, libraries and other products that are used by a high percentage of fabless companies. Instead of being "outsource" suppliers who implement existing technology, the large IP suppliers will become industry technology drivers, combining fundamental technology development with efficient productization.

The smaller "boutique" IP companies will still maintain an important role. These companies will blend consulting, design services and IP creation to provide highly tailored niche products to customers' precise needs and applications. These boutique suppliers will be the choice for outsourcing development of existing technology, with their primary value as product implementers. They will also serve as alternatives to the dominant IP suppliers, thus putting significant pressure on the leaders to continually innovate and improve their products.

7.11.2 Impact on the Fabless Industry

The maturation of the IP industry will be a boon to the fabless community. Technology content will increase, choices will expand and prices will drop. The fabless community will leverage these shifts to increase their technology differentiation while maintaining cost leadership.

But the biggest advantage to the fabless community will be the increased variety of available leading technologies. The category-leading IP suppliers are rapidly becoming the technology drivers in their respective areas. Due to their leverage, they can invest in fundamental research, EDA tool infrastructure, rigorous quality control and other product aspects that are beyond the reach of any other companies investing in IP. Even the largest IDMs are procuring IP from commercial sources rather than developing it internally. With the same IP going to both the fabless and IDM markets, there will be no "IP gap" between fabless and IDM design communities.

Pure-play foundries will see most of the same advantages as the fabless designers. Widely available, high-value IP will enable the foundries to compete very effectively for the most advanced systems-on-a-chip (SOC) designs. One additional advantage will be in manufacturing cost savings. The quality of physical IP has a direct impact on the yield and manufacturability of a chip. IP bugs result in production halts, the worst fear of both the foundry and the chip designer.

Large IP suppliers will invest enormously in rigorous, high-quality product development flows, since the expense can be amortized across hundreds of uses. Similarly, the products will be proven in silicon hundreds of times. Proliferation will directly lead to exceptionally high-quality IP.

7.12 Intellectual Property Considerations

7.12.1 Background

Initially the semiconductor suppliers were vertically integrated. They were required to develop and support all of the elements necessary to develop and manufacture semiconductor products. This included everything from process and library development to design tools, manufacturing and distribution. The investment and infrastructure required for a vertically integrated company was prohibitively expensive and limited access to the semiconductor industry to very large multi-national companies. As the industry began to mature, however, it began to disaggregate as companies started to specialize in specific aspects of semiconductor development and manufacturing (Figure 7.15).

Three key industries, which arose from this desegregation, in turn reduced the barriers of entry to the semiconductor industry, and made it possible for smaller companies to compete. The

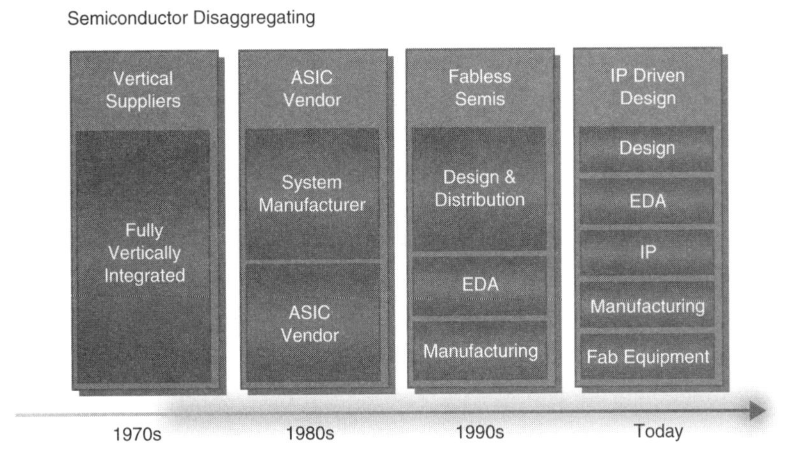

Figure 7.15

EDA, independent third-party foundries and the SIP industry, provided access to the design and manufacturing infrastructure previously limited to large companies. The EDA industry provided designers ready access to sophisticated design tools and environments which allowed them to effectively design and test their products. The foundries provided small companies access to cost-effective semiconductor manufacturing without incurring the expense of the development of manufacturing facilities and infrastructure.

SIP suppliers provided designers quick easy silicon-proven access to standard's based IP which allowed the designers to reduce their risk and effort associated with the development of the standards portion of their design, and focus their efforts on the portions of the design which allowed them to differentiate themselves in the market place. ARM for instance, an early entrant into the SIP industry, provided silicon-proven embedded CPU technology to companies, which did not have the expertise or the capital to develop the infrastructure to develop and support their own CPU architecture.

Prior to ARM, most semiconductor companies developed their technology hoping to obtain a return-on-investment (ROI) through chip sales. By providing their own CPU technology they were able to differentiate their products, and lock their customers into their products. ARM decided that rather than product silicon, it would obtain the return on its investment through licensing designs to others, and thus obtain revenues through license fees and royalties associated with the production of their technology. This not only provided small fabless companies easy access to proven cutting-edge CPU technology, it also provided the industry with the ability to create products based on industry standard IP, and to have third parties develop supporting solutions and products to support and complement the IP – two benefits which cannot be matched by internally developed SIP.

There is a constant push toward cost reduction, which is accomplished by greater and greater levels of integration. As the new smaller geometries come on-line, designers are able to put additional functionality in the same die area, this allows the customer to reduce the number of components in their system and therefore reduce their overall bill of materials. This constant march towards miniaturization requires the designer to provide ever increasing levels of functionality, and increased gate count into their design. Moore's law says that technology advances to the next generation every 18 months. This effectively doubles the density and the number of gates which fill in the same area as the previous process. As a result, it is difficult for any design team to fill the available area on a chip. To meet the market window, with available design resources, and cost constraints, designers must obtain proven third-party IP to fill the ever increasing chip gate count.

7.12.2 IP Business Model

SIP companies are similar in many ways to the fabless customers they support. It goes without saying that they have R&D investment, investors to satisfy, and mouths to feed at home. They are under the same cost, revenue targets and ROI calculation pressures that their customers are under. The only real difference is that the fabless company obtains its revenues and ROI through sales of physical chips. An SIP company receives its ROI; by spreading its development cost across multiply licensees and royalties. As a result, IP providers usually target standards-based or fairly broad market opportunities, as they need to spread their investment across multiple license engagements. They must differentiate amongst other similarly situated IP providers, while still addressing a large enough market to receive a reasonable return on their investment. This necessitates that they either develop standards-based solutions or develop IP which they then drive to become industry standard (or prominent) through their licensing activities.

As a result SIP customers should look at their IP licensing activities as a method of obtaining quick proven SIP for their designs, but not necessarily to differentiate their product. Customers need to differentiate through the features they add to the design and the manner in which they architect the overall design. Designers should not look to the SIP to be the sole differentiator in their design. There is an abundant supply of standards-based IP. The IP model of spreading the development expense across multiple customers and market forces mean that the acquisition price is usually below the cost of developing similar IP, so designers should license available IP and focus their efforts on chip design which allows them to differentiate their product beyond the SIP included in the chip.

The primary benefits of the IP model are reduction of effort and expertise, time to market and risk reduction. Often, the actual designs may not be terrible sophisticated, and the designer may be able to reproduce a similar solution, but the fact that the IP vendor provides a proven, supported off-the-shelf product can be the difference between a successful design, and product launch, and a run-away project schedule. As a result, the main value IP providers are providing is not the design itself, but the time-to-market, and the reduction of effort and risk in the project design. As mentioned earlier IP providers, EDA companies and foundries provide very small companies the ability to create very complex designs, containing ever increasing functionality, at competitive prices to their customers. These three industries provide the foundation for the fabless business model and allow designers to compete in the competitive semiconductor market with relatively little investment or infrastructure.

7.12.3 Licensing

There are two main primary components to the licensing revenue, (1) the license or access fee, this provides access to the SIP, and provides the rights necessary to design products using the SIP in a design and (2) royalties, which are the compensation for the right to manufacture using the SIP.

The licensing or access fee can be split using a number of different business models. These can be fully paid where there is simply an initial license fee for access to the IP. This access can also be split across delivery milestones if the SIP requires some customization or re-characterizations for the customer's requirements, or the license fee can be amortized across the customer's uses, by paying a fee for each design which integrates the SIP.

7.12.4 IP Delivery Mode

SIP can be provided in two different forms, (1) synthesizable RTL or "Soft IP;" this is a high-level descriptive design language which must be linked to a process library prior to use and (2) GDSII, or "Hard IP." GDSII is SIP which has already been targeted or "hardened" for a specific process and library.

There are a few primary differences between the two types of IP, and both have their advantages. Soft IP provides the designer flexibility. They can target a specific foundry and process variant, by targeting the RTL to a specific process library. This provides the ability to target designs towards, power, performance and area (PPA) through the use of a process and process library which have desired characteristics. Many forms of SIP also have configuration options. Many soft CPUs for instance have scalable caches, which allow the designer to use the optimal cache configuration for their application. PCI express and other high end SIP I/O standards allow the customer to identify the number of channels to be included in the design, and to turn certain features on and off, based on product needs.

Soft IP provides the designer flexibility to tune their design to application requirements, but also requires additional time, effort, expertise and risk. GDSII, or hard IP, on the other hand, provides a silicon-proven solution which has been targeted to a specific foundry and process. This reduces the level of effort and risk associated with the design, but also limits the designer's ability to tune the design to fit their specific needs. Hard IP is preconfigured in an identified block, targeted at a specific process, and implementing a specific library, as a result the designer's options are limited, but specific implementation has most likely been proven in silicon, which reduces the designers design risk, and the complexity and the effort associated with designing with the given IP block. The cell can simply be placed and routed into the design.

The risk of using RTL can be mitigated through the development of reference methodologies, a vendor-developed design environment and flow which has been validated through a working test chip, but each implementation of RTL is unique and therefore subject to greater risk than the integration of silicon-proven GDSII. GDSII freezes all of the variables in the SIP design of the block which is validated in a test chip. It therefore provides the lowest risk avenue towards chip integration of IP.

Analog IP is very process dependant, and is very difficult to design and produce; as a result IP blocks which include analog IP are usually distributed as hardened IP to increase the likelihood of a working solution and to reduce the design complexity and risk (Figure 7.16).

	Synthesizable IP	GDSII
Flexibility in design	High	Low
Design effort required	High	Low
Design experience needed	High	Low
Design risk	Moderate	Low
Manufacturing flexibility	High	Low
Source: Artisan		

Figure 7.16

7.12.5 Quality of IP

Every project begins with a technical specification, which spawns a discussion of "make versus Buy" on each of the functional blocks. For many of the blocks the choice is obvious. The design team may not have the time, resources or experience to do the design internally, it may also require standards-based IP which, in the case of a CPU may only be available from one company.

Each additional block which is developed internally adds additional risk to the design. If the designer uses pre-validated blocks, they are able to reduce the overall risk of the project.

Reuse has been a buzz word in the design community for several years. Reuse dictates that you design in such a manner that the IP block be easy to reuse in the next project, thus reducing the overall cost of the block from design to design. Unfortunately, most internal projects are very focused on the development and tapeout schedule of the current project, and do not have the bandwidth or luxury to focus on the documentation, design, and design methodology necessary to enable reuse.

SIP providers, on the other hand, are in the business of enabling reuse. To achieve success and the return on their investment, they must provide a robust design, with sufficient documentation and support to (1) license their SIP to a number of customers, (2) enable a number of designs and (3) eventually receive royalties from the production of their SIP in customer's end products. Some SIP vendors have a better grasp of this, and are more successful than others.

Successful SIP vendors focus on getting the customers successfully to silicon. This includes adding a test chip "silicon validation" to the development schedule to assure that the final design has actually been manufactured and tested. This should provide valuable feedback to the SIP provider as to any bugs contained in the design, and the manufacturability of the design, and should provide confidence to the customer that they are designing in a proven solution. In fact, due to the business model, SIP is validated through multiple customers, multiple designs and more production volume that independently developed solutions.

By licensing IP, the designer does not reduce all risk in the chip, there is still a risk that the design might not work in their specific implementation, and if this is the case they are at the mercy of the IP vendors support organization to debug any problems in the licensed IP.

7.13 IP Outsourcing

Designers are facing increased market pressure to rapidly introduce new products, which shortens the time available for research and development.

Many semiconductor companies, both fabbed and fabless, are increasingly relying on external sources or technical expertise for various components of the SOC design. The use of proven third-party SIP components allows semiconductor companies to meet market pressures while continuing to focus on the portions of the SOC that constitute their core competencies.

The independent third-party IP industry was worth approximately $2 billion in 2006 and is growing at 20 percent annually. By 2010, IP revenues are expected to double.

The use of third-party IP facilitates design reuse and allows for a building-block design approach. With silicon complexity increasing at more than twice the rate of design productivity, a significant "design gap" has emerged, as illustrated in Figure 7.17. To meet

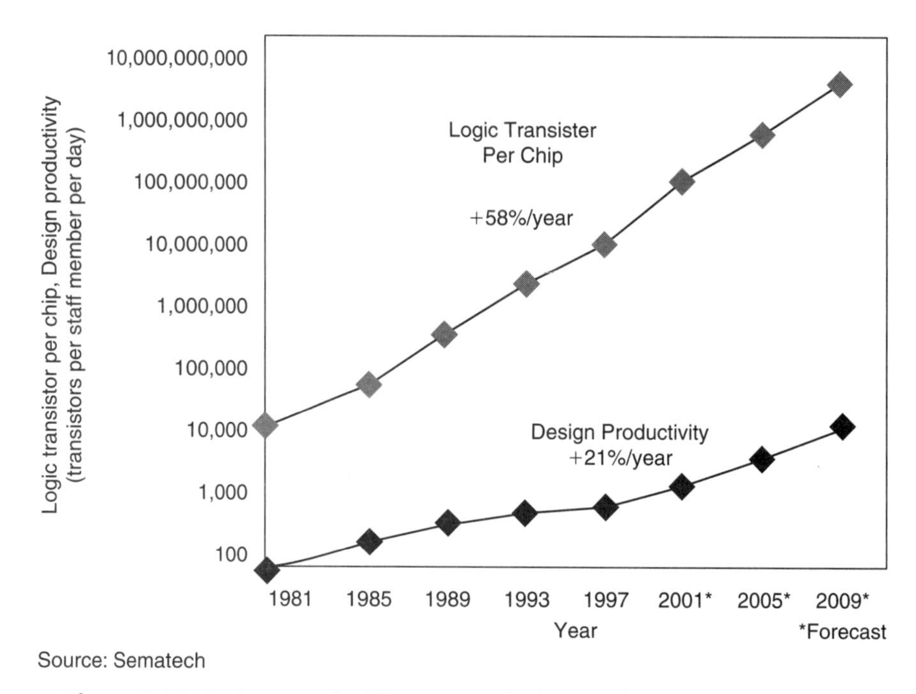

Figure 7.17 An Increase in Silicon Complexity Results in a "Design Gap"

ever-shrinking design cycles, semiconductor and electronic systems companies are turning to SIP suppliers for pre-designed, proven-in-silicon circuit elements such as microprocessors, analog or mixed-signal, general-purpose logic and memory blocks.

One area in which fabless companies can really take advantage of third-party IP is embedded memory, which is picking up momentum due to the rise of the Internet and communications. In fact, over half of the chip's surface will be memory (Figure 7.18).

Until now, system designers have generally used stand-alone or discrete memories. The transition from stand-alone to embedded memory is occurring because of the growth of the Internet and the development of the optical and wireless communications infrastructures, which create demand for communications equipment and digital appliances. The system designers creating these products are seeking technologies that will decrease the size and enhance the performance of their products. Embedded memory, which has become the choice of many designers over traditional stand-alone memories, is driving the explosion as the demand for bandwidth continues to grow. In fact, for every 10x increase in bandwidth demand there is a 5x-density increase for memories. Leading SIP providers are focusing on embedded memory with an emphasis on manufacturability to meet the challenges of even higher density memories used in future SOCs.

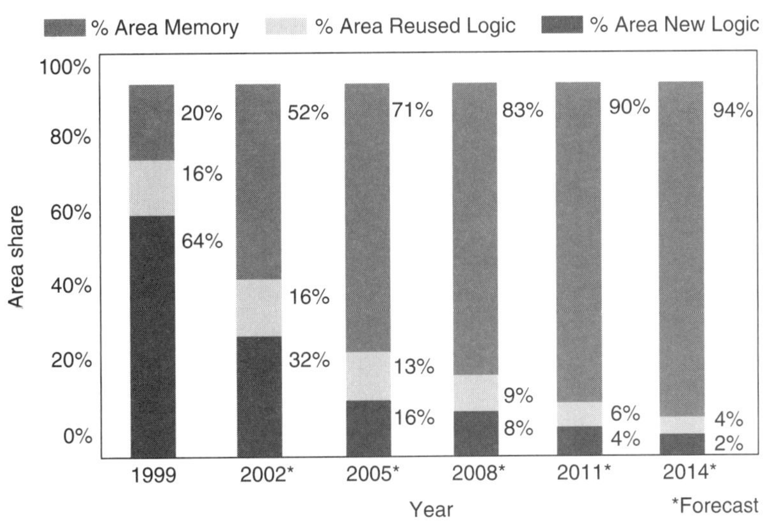

Figure 7.18 Embedded Memory will Take Over Chip Surface

7.13.1 The Need for Reliable Memory IP

When custom design techniques are used, embedded memories can be superior in terms of size and performance as stand-alone memories. Designing application-specific memories used in SOCs requires special knowledge and expertise. Furthermore, the architecture of a specific memory must be redesigned and re-characterized for each new generation of process technology, as well as for each different foundry, which is a time-consuming process.

The internal design teams of semiconductor companies typically lack the dedicated resources necessary to keep pace with rapidly evolving memory designs. Suppliers of stand-alone memories for personal computers or other devices that include memory as a separate element on a circuit board, as well as suppliers of other SOC components, often lack the design expertise and software tools necessary to provide the wide range of embedded memories used in SOC designs. These factors have created a market need for highly reliable, application-specific embedded memory IP.

Fabless semiconductor companies are purchasing embedded memories from SIP providers for products that utilize SOC technologies with large memory requirements and high-performance, low-power architectures. These products include:

- *Communications and Internet Infrastructure*: Communications SOCs are used throughout the Internet and are found in routers, switches, digital subscriber line (DSL) modems, gigabit Ethernet equipment and home networking. For these applications, the quickly accelerating demand for bandwidth is driving the need for increased memory capacity and faster memories.

- *Consumer Appliances*: Internet appliances and other consumer products increasingly require greater functionality, Internet connectivity and low-power consumption.
- *Computers*: Computation equipment such as personal computers, workstations and servers require more complex chip sets and high-performance embedded memory to achieve new features such as advanced 3D graphics and DSP.
- *Communications and Internet Infrastructure*: Communications SOCs are used throughout the Internet and are found in routers, switches, DSL modems, gigabit Ethernet equipment and home networking. For these applications, the quickly accelerating demand for bandwidth is driving the need for increased memory capacity and faster memories.
- *Consumer Appliances*: Internet appliances and other consumer products increasingly require greater functionality, Internet connectivity and low-power consumption.

SOC designers must have absolute confidence in the design and manufacturability of the embedded memory used in their designs. This requires a robust design approach together with a comprehensive silicon validation program to ensure that memories will meet specifications and high yield. Memories must be manufactured, tested and characterized before being integrated into products. Furthermore, each design should be optimized across multiple foundries to allow for an uninterrupted source of supply and to secure low prices. By purchasing memory IP that has already been tested and characterized, fabless vendors are able to shave months off their product cycle times with the assurance that these memories will work as designed.

Specializing in design expertise, these SIP companies can offer their fabless customers leading-edge technologies for advanced manufacturing processes. For example, Virage Logic has assembled a global team dedicated exclusively to memory design. This team includes senior-level engineers with significant expertise in various types of memory design, including SRAM, Specialty RAMs and DRAM.

While the embedded microprocessor and DSP cores are essential in defining the system architecture, embedded memory is key to ensuring design manufacturability at cost-effective levels. With larger memories being integrated onto the SOC, the embedded memory – not the embedded microprocessor – determines the manufacturability and cost of the SOC, as well as how quickly time-to-volume is achieved. Memory is a magnet for defects during IC manufacturing because it typically has twice the wafer defect density of logic.

7.13.2 Embedding Multiple Megabits Cost-Effectively

A way to combat some of the reliability issues during manufacturing is to ensure that the design is verified in silicon, which will guarantee silicon performance. Built-in self-test (BIST), built-in self-repair (BISR) and diagnostics guarantee high quality and yield. Widespread support for leading foundries assures low prices and multiple sources, which has allowed embedded memory compilers to support all types of memories.

Achieving high yield on embedded memories that are 1 megabyte or more requires a manufacturable design together with redundancy and a test and repair capability. The resulting memory allows multiple-megabyte of SRAM to be embedded cost-effectively on a sub 0.13-micron process. Redundancy is commonly used in stand-alone memories of 1 megabyte or larger to make them repairable. As memories become larger, they are more expensive to produce, and at the same time, are more prone to defects. Redundancy provides for additional bits to be designed into the memory to replace defective bits, which helps increase yield and reduce cost. This same approach is used with embedded memories.

Commercial embedded memory suppliers now have the breadth of product line and expertise to ensure fabless vendors that by outsourcing their memory IP they can save time and money, and get a higher quality product. For more companies to consider outsourcing their memory IP to third-party vendors, it will require embedded memory to provide additional value over stand-alone memory solutions. Embedded memory achieves this without requiring the system designer to be a memory designer. With a complete set of validated EDA views, integration into the design reuse methodology is assured. These factors have created a market need for third-party providers of highly reliable, high-performance embedded memory IP for SOC designs.

7.14 Making IP Work in the Fabless Semiconductor Community

As chip design and manufacturing complexities continue to increase exponentially, it is becoming less practical for fabless companies to internally develop the necessary semiconductor IP needed for SOC designs while meeting critical market windows to stay competitive. There's no question that licensing IP from third parties, rather than building it internally, saves critical cycle time and has become an increasingly important differentiator. IP blocks available from IP providers are becoming more and more complex and therefore more and more difficult to integrate and support. There are a number of issues associated with the use of third-party IP in SOC designs, but two issues facing the fabless community have bubbled to the top of the priority list. The first is IP quality and the second is IP portability.

7.14.1 The Importance of Third-Party IP in SOC Design

An analysis of current and future IC designs demonstrates the increasing importance of third-party IP integration and support as feature sizes are reduced to 130-nanometer, 90-nanometer and below. With IC designs becoming more and more complex, an ever-growing percentage of transistors are becoming increasingly IP-centric. As a result, access to the necessary IP portfolios is crucial, as is the ability to efficiently utilize the IP within the designs. As design integration levels increase, the role of mixed-signal, memory and processor IP becomes more significant.

IP concepts within the IC industry are maturing and the various approaches that IC vendors have taken to accessing and using IP have already changed significantly. A common misconception among IC vendors is that IP would be available from multiple sources and that IC vendors would be able to license IP based on the requirements of specific designs. IP was expected to be synthesizable, with seamless integration into designs. This approach has not been successful because IP available from third-party vendors is often incomplete or alternatively fine-tuned for specific wafer processes. Additionally, synthesizable IP may only be effective for low-complexity designs.

Third-party IP has not yet been widely adopted, resulting in a slower-than-expected growth rate for stand-alone IP vendors. Various IP company business models have not always supported seamless access to IP, thus preventing IC vendors from bringing new designs to market within shorter time windows. In many cases, third-party IP is not thoroughly characterized. Synthesizable IP has been ineffective, and the cost of IP characterization is often comparable to the cost of internally developing IP. An increasing percentage of IP involves mixed-signal functionality, in which the IP must be characterized at the transistor level within the targeted process.

7.14.2 IP Quality

The most significant challenge associated with using externally purchased IP is ensuring the IP block's quality. When a third-party provider is selling IP, expectations are that it should work without "bugs." However, this may not often be the case. In many instances, quality does not meet expectations. If "bugs" are identified, the corresponding remedies may not always be incorporated into the IP vendor's database. There continues to be a major tradeoff that IP vendors must manage between making IP available as soon as possible (for revenue-generation purposes), and releasing it before it has been verified to have a high degree of quality. Inadequate IP quality has too often been the end result of this tradeoff, resulting in a major roadblock to the successful adoption of third-party IP.

Is it possible for a set of tools be developed, endorsed and adopted to facilitate the inclusion of third-party IP into the fabless semiconductor supply chain? Similarly, can ISO-9000-like verification criteria be defined to insure the quality of IP similar to criteria used by other members of the fabless supply chain such as foundries, packaging houses and test houses?

To answer many of these questions, FSA's IP Subcommittee commissioned two working groups – one to focus on hard IP and licensing issues. To clarify, hard IP is process-specific and is generally delivered in GDSII format. It is optimized for power, size or performance and is mapped to a specific technology. The goal of each of these working groups is to further clarify these issues and investigate and leverage existing industry efforts on both fronts. The intention is not to duplicate current efforts, but rather to accelerate, extend and implement them.

7.14.3 IP Portability

One goal of the fabless community has been to eliminate barriers for designers considering the COT model. The benefits that the COT model can bring designers include such things as (1) more control over the design by the designer and the ability to exert that control with multiple manufacturing partners (i.e. foundries), (2) more flexibility and freedom of choice as it pertains to IP suppliers, foundries, packaging and test houses and (3) the ability to take on more risk with regard to timing closure and the delivery of first pass working silicon. COT customers want the flexibility to bring their design to their foundry of choice and the subsequent freedom to move it to another foundry later, if they so desire. Customers also want to insure supply by having designs manufactured at more than one foundry. Although this is not a concern in the current environment, this issue becomes increasingly important when foundry allocations become scarcer.

Regarding IP portability, outsourcing is here to stay. The health and competitiveness of fabless semiconductor companies is dependent upon freedom of choice, and second sourcing has become increasingly important. To that end, IP portability is a major obstacle that most fabless companies are simply unaware of, even if their designs contain all digital logic, Artisan libraries at TSMC or UMC contain bit cells that are foundry specific and are, therefore, not easily portable to other foundries.

The best way to address the problem would be for the industry to adopt a "standard library" concept. The benefits of such a library include footprint compatibility, reduction in re-design efforts, reduction in time to tapeout and overall reduction in design porting risk. The caveats of such an approach, however, are that it may not be a cure-all as analog, radio frequency (RF), specialty blocks and some other IP will still require porting and verification. Intrinsic and interconnect timing (delays) will not be the same due to process differences. An "Industry Baseline" bit cell is also required, and finally, consideration must be made for potential performance degradation.

7.14.4 Conclusion

The semiconductor industry continues to challenge the electronic design community by making available silicon capacity that exceeds by far what today's designers can utilize in a reasonable amount of time. IP represents a critical capability for effective participation in the fabless semiconductor community as well as the semiconductor market as a whole. A number of issues exist in the IP space that will continue to make its use in SOC designs more and more difficult if not addressed promptly and properly.

7.15 IP Acquisition Considerations for Fabless IC Companies

In today's changing foundry environment, the emergence of increasing numbers of manufacturing foundries has opened many opportunities for IC designers, developers and IP providers. Previously, customers or companies that needed manufacturing sources needed large capital to build their own manufacturing facilities, or only had several foundry choices in Asia and the United States. But with the new crop of foundries stabilizing production in areas worldwide from China to Malaysia to Singapore to Israel, IC developers are finding more low-cost solutions at their fingertips, and the semiconductor supply chain has become increasingly stratified (Figure 7.19).

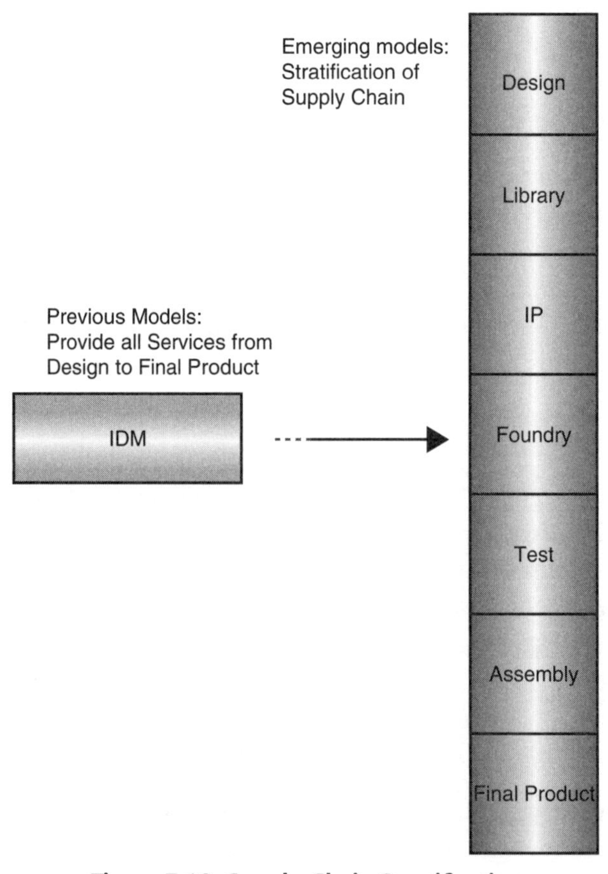

Figure 7.19 Supply-Chain Stratification

To compete against foundries that have had many years of silicon production, these foundries have branched off to focus on various areas of IC manufacturing, whether in low-cost wafers, IP solutions, consumer process technologies, embedded memory or stand-alone memory. To compete, many of these foundries need to offer low-cost wafers to their customers as they ramp up their services and production. This results in various ramifications to foundry customers and strategic partners, particularly third-party IP providers and designers.

Previously, foundries often had to maintain costly internal design teams to support customers with IP and design for ICs manufactured at the foundry site. Because wafer prices could remain at a premium due to the lack of alternative manufacturing sources, this strategy was feasible. However, with increased competition, maintaining an internal design group no longer makes financial or technical sense for foundries. Not only is it too expensive to maintain an internal design support team (when competing on lower wafer pricing), but it is also very difficult to find staff that has experience in the wide range of IPs that a foundry will offer. Foundries have always had various strategic IP partners for logic, analog, memory, etc. These growing opportunities for IP providers and designers have also given way to various modes of operation and business models to sustain the growth of these fabless IP companies, their partner foundries and their customers.

Foundries benefit from outsourcing their design services because they do not have to bear the overhead and costs of maintaining a design group. They also relieve themselves of the responsibility of guaranteeing IP and design quality, which is outside of their core competency. There are less time constraints on staff as well, as IP partners can directly interface and deal with customers who are licensing their IP. Lastly, since the foundry can maintain many design and IP alliances, they can partner with companies focused on their specific target market and have the highest design experience for their particular product or service. In particular, for non-standard designs such as embedded memory for ASICs, high levels of technical expertise and time are necessary to develop designs and IP that are cost-effective and maximize performance and operating ranges.

Customers benefit because they receive competitive wafer pricing paired with premium design IP and service. However, there are still some strategies and considerations that IC developers need to consider to successfully acquire third-party IP to benefit both short- and long-term goals.

7.15.1 IP Strategies for IC Companies

For IC companies to successfully take advantage of the available third-party IPs, there are several things to consider in reducing the time, money and resources involved in acquiring IP.

The importance of reusing IP in today's increasingly stratified design chain is to shorten the product development time, as well as to reduce the risk and costs involved in developing new products. Particularly as process technologies shrink, chip design becomes more and more complicated. Reusing IP that has already been proven, or is from an IP provider focused on certain segments, can mitigate some of the risk involved in designing an entire chip from scratch internally. In addition, because IC design houses may be more focused on their own CPU, logic or analog design, acquiring IP from various parties may enable them to create higher quality designs than if they had tried to do the entire design on their own.

There are several strategies to contemplate in acquiring IP, depending on a company's technology roadmap, financial resources and design capabilities. Depending on the foundry partner and the type of IP required, costs can range from free of charge to upwards of hundreds of thousands of dollars (Figure 7.20).

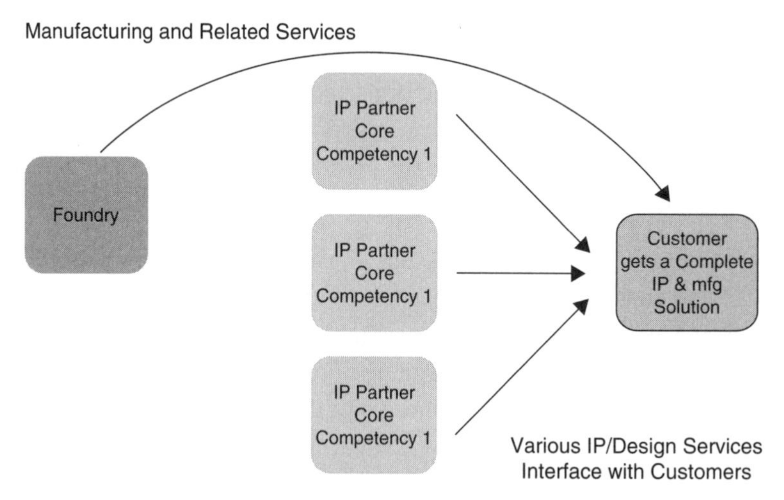

Figure 7.20 Foundry–IP–Customer Relationship

7.15.2 Technology Roadmap

If a company has a technology roadmap based on several key IPs, it may be beneficial in the long term to fully acquire the necessary IP to sustain its technology and product growth. This type of outright IP acquisition, which usually includes database transfers, netlists, etc., as opposed to black box licensing, often has higher associated up-front costs, such as NREs. Sometimes IP companies will be flexible and can also offer lower NREs with some combination of recurring payments or royalties. However, the terms of payment should depend on the type of revenue stream that can be generated from the various products, and whether it makes sense to acquire the IP outright or to license it. Also, if the product roadmap for that company is dependent on just a few key IPs, it makes more sense to own the IP and have the flexibility to use it wherever it is needed, rather than build up a large royalty expense. In addition, if the company has the design capability to be able to support and improve upon the acquired IP, it can eliminate the need to purchase more IP for other products further down the road.

7.15.3 Financial Resources

The amount of financial resources available for a company to invest in R&D is a large factor in the types of licensing agreements it will sign with its providers. If a company has the cash up-front to be able to pay for an IP block, it may make more sense as royalty or recurring fees may add up to larger costs in the long term. This needs a calculation of the expected volume production of the product(s) using the IP, and a comparison of that versus full up-front licensing fees.

7.15.4 Design Capabilities

The design capabilities are also a major determining factor in a company's decision to outsource IP. Obviously, if the company does not have the design knowledge, acquiring third-party IP is a must. However, there are often in-between stages in which companies do have the capability to complete designs internally. But in these cases, there are often advantages to outsourcing IP anyway. For example, Company A may have the ability to internally do a design. However, there is also available IP from Company B that can be acquired. Instead of reinventing the wheel, which may become costly and unpredictable, it may make more sense for Company A to acquire IP from an outside party. By licensing a silicon-proven IP design, Company A reduces its risk of failure, reduces its product development time and improves its performance (e.g. higher yields, increased functions, design flexibility, etc.). The importance is to find a trustworthy, experienced partner that can guarantee a working design, as well as product security and a range of designs and foundry partners for manufacturing flexibility.

As can be seen, there are many considerations that IC companies need to take into consideration for successful navigation through the fabless IP market. The number of companies utilizing third-party IP is increasing, as is the number of IP providers. Through a careful IP strategy balancing time, money and technical capability, IC companies can optimize their design chains to come out with quality, cost-effective products.

CHAPTER 8

e-Commerce

8.1 The Virtual Fab Challenge

In the outsourced environment, increasing competition and disparate operations brings forward new sets of challenges in ensuring the accuracy, profitability and timeliness of the fabless organization. The foremost challenge arising in the fabless manufacturing environment is the ability to manage trading partners and subcontract manufacturers and still maintain real-time inventory visibility.

Much of the traditional manufacturing has been done within a single factory, which is managed and owned by the same organization. In the fabless model, due to outsourcing of wafer fabrication, assembly and test operations, the fabless company operates in a virtual fab environment which relies on information that is being provided by their trading partners, therefore placing an undue overhead in monitoring customer on-time delivery. To truly address the outsourcing challenge fabless companies need to embrace a methodology of being able to achieve up-to-date and accurate inventory visibility with their suppliers and then make intelligent business decisions converting data into information. Once a fabless company can establish the inventory visibility, the next challenge that surfaces quickly is being able to manage order fulfillment to their customers while using third-party shipment services to drop-ship finished products. This becomes significantly harder in the context of using distributors and contract manufacturers.

The fabless model has challenges with respect to reconciling the financials; typically vertically integrated manufacturers only deal with cost of production, capital and cost of goods issues. However, in the fabless environment, the financials have to take into account standard and actual costs due to yield variances, purchase price variance against a blanket purchase agreement with fab/assembly houses and, above all, reconciliation of distributor credits, charge backs, reworks and scrap. Further, with decreasing electronic prices the importance of managing cost variance becomes crucial in affecting the fabless company's bottom line and therefore its profitability.

As for the fabless companies that are fortunate enough to be on a path to a public offering, they are likely to face new sets of challenges with increased need for financial compliance, stipulated under the Sarbanes Oxley Act. The first problem auditors face with respect to compliance is in a physical audit of the inventory which now resides with outsourced partners. The act of reporting and the process of collecting the inventory information from outsourced partners are of paramount importance in a smooth transition from a private to a public company.

Most of these challenges need to be addressed as part of any information system the fabless companies embrace. However with decreasing margins, the systems need to contribute to a higher return on investment and this can be achieved by lowering the overall financial exposure of the software systems. With the advent of software as a service (SaaS), companies can now benefit from choosing software vendors that offer monthly subscription fees for use, thereby reducing the overall cost of ownership.

8.2 Semiconductor & Fabless Manufacturing: What is Different?

8.2.1 The Inverted Bill-of-Material: Discrete versus Process

Much of the world's manufacturing is described as discrete oriented. Many individual parts or components are built into one or more levels of semi-finished or finished goods. The bill-of-material (BOM) for this type of manufacturing is a "many-to-one" hierarchy.

Integrated circuit (IC) manufacturing is described as process-flow-oriented. An individual material, a wafer, is processed with a set of photomasks to make various types of chips. These chips can be sorted (tested) into categories with each category having different electrical characteristics. Chips from a category can then be assembled into a variety of different package types making, a series of unique chip/package combinations. These packaged chips can then be tested into a variety of finished products based on results at a final test operation. The BOM for this type of manufacturing is a "one-to-many" hierarchy and is usually referred to as an inverted or reverse BOM.

One of the major challenges of discrete manufacturing is the management of a potentially large BOM; therefore, early manufacturing support software concentrated on this as critical and is categorized as Materials Requirement Planning (MRP) software. As the complexities of manufacturing grew and the sophistication of software increased, MRP software began to add functionality to manage other manufacturing resources such as labor and equipment. In some cases, support for purchasing and financial reporting was included. The acronym was upgraded to MRP II and the words were changed to Manufacturing Resource Planning. As manufacturing expanded beyond the walls of a single factory and took on attributes of a regional or global supply chain, the name was changed to Enterprise Resource Planning (ERP) and included full financials, purchasing, order management, distribution, logistics, etc. Additional functions such as Product Data Management (PDM), Customer Relationship

Management (CRM) and Sales Force Automation (SFA) continue to be either added-in or bolted onto ERP systems.

ERP software is sometimes classed into tiers, depending on functionality. Tier 1 ERP systems are rich in functionality and are capable of managing large, globally distributed manufacturing companies. Examples of Tier 1 systems used in the semiconductor industry are Baan, IFS, JD Edwards, Oracle, PeopleSoft and SAP.

Tier 2 packages may have less functionality or are based on older underlying technology and include suppliers like Great Plains and NaVision.

Tier 3 packages have limited functionality and may look much like an MRP II package. Suppliers that have sold into the semiconductor industry include Expandable, QAD and Ramco.

Semiconductor companies that plan to grow rapidly and want broad functionality are attracted to ERP systems because they include support for the major business processes including finance (A/R, A/P, G/L, F/A, etc.), order management, inventory, purchasing, project management, project costing and manufacturing. Because ERP packages grew out of MRP II functionality, they are typically oriented toward discrete manufacturing, however, and have limited or no support for semiconductor process manufacturing. Most semiconductor companies that purchase ERP systems make little use of the MRP module or Master Scheduling (MS) module, as they do not function properly for inverted BOM style manufacturing.

Other functions that are included in ERP packages may not function properly for semiconductor companies because they are designed to work for discrete manufacturing and not for semiconductors with an inverted BOM structure. As an example, product costing for discrete manufacturing is typically built from the bottom up with costs being arithmetically accumulated through the various levels in the master BOM until reaching the finished product level. For semiconductor products, the effective cost of the die or other inputs to cost, such as a leadframe, continues to go up as the product goes through operations where there is fallout or scrap. This yielded cost approach requires the system to store the inverted bill-of-materials (IBOM) structure and use a different costing algorithm from that for discrete manufacturing. Typical ERP Available-to-Promise (ATP) functionality suffers a similar fate for the same reasons. These are but two examples of ERP functionality that do not work properly because of lack of support for the IBOM structure and lack of process manufacturing-oriented algorithms.

Many semiconductor companies or larger companies with semiconductor divisions also do discrete manufacturing. Examples are fabless companies that also sell board-level products (e.g. ATI Technologies (now part of AMD) or Adaptec), multi-product companies (e.g.

Motorola or IBM) and semiconductor companies that supply evaluation/reference boards or multichip modules. The discrete manufacturing operations within these companies can use the standard ERP manufacturing functionality.

Virtually all semiconductor companies, whether fabless or integrated device manufacturer (IDM), need support for managing their manufacturing operations. Larger companies, particularly those that do some or all their own manufacturing have Manufacturing Execution Systems (MES) to run their wafer fab, probe, assembly and test operations. Examples of software include Promis, FASTech, Consilium, Camstar and SI View. These packages were designed from the ground up to support semiconductor process flow manufacturing. Companies that have MES packages installed typically use them to manage not only their internal manufacturing but also their outsourced manufacturing.

Most fabless companies, whether they have an ERP system or simple accounting system installed, will not have an MES system installed, but will manage their manufacturing with some combination of Excel spreadsheets and Access databases. In the past couple of years, a new class of software called "Fabless Supply Chain Management" or "Fabless SCM" has begun to evolve. Examples are Oracle "OSFM," TenSoft's "Fabless Semiconductor Management," and Serus. These packages vary in functionality, but generally include IBOM support for work-in-process (WIP) tracking and lot genealogy, as well as things such as order management, PO/SO tracking, ATP, engineering data storage, yield reporting, etc. In all cases, they offer mechanisms to get data from the manufacturing partners into their system. Some of these have been integrated with ERP packages such as Oracle, Great Plains and NaVision.

8.2.2 Multiple Routings

Production lots for a particular product may be routed through different combinations of factories on an ad hoc basis for various reasons, (cost, cycle time, customer qualification, yield, etc.). It is important to be able to keep up with what factories worked on a production lot during its manufacturing life. As some customers will only accept product built – in a particular factory, with a particular process, with particular equipment or a combination – it is important to be able to manage this situation. When a customer places this type of order, care must be taken to build the product using the right resources. When it comes to shipment, the company must ensure that the product received the right processing for that customer.

8.2.3 Splits/Recombines/Binning

For various reasons, production lots may be split into multiple sub-lots for some of their processing. These sub-lots may stay separated until completion or may be combined back together. Portions of lots or whole lots may be scrapped due to processing issues. At some later time, often after some rework, this material may "reappear" as a bonus lot. Lots may go through a binning operation, usually at test but sometimes at probe, where the starting product becomes one or more new products based on electrical characteristics such as speed or temperature characteristics.

8.2.4 Wafer to Die Conversions

The first phase of IC manufacturing is wafer fab. Silicon wafers go through a series of processes that build an array of chips into and on top of the surface of the wafer. Depending on wafer diameter and chip size, there may be dozens to thousands of chips on each wafer. This constant number of chips on a wafer is known as potential good die per wafer. During the fab process, it is unknown which of the die or chips are good and which are defective. These two units of measure must be comprehended in systems that handle manufacturing information.

8.2.5 Yield

The major process areas in IC manufacturing are fab, probe, assembly and test. It is unusual to experience major yield losses in the actual fab or assembly operations. If the fab production lot has 24 wafers at initiation, it usually will have 24 wafers at completion of the process. This does not mean that all chips are good, only that all the wafers in the lot made it through without being broken or otherwise scrapped. This also does not mean that data from the fab operation is not important. It is this in-process data from the fab that helps companies understand why some chips are defective at probe. Likewise, usually a small percentage (<1.0 percent) of units in assembly are scrapped in the process itself. Most yield loss in the "back-end" (assembly and test operations) occurs in test. This yield loss includes chips that were marginal at probe or include defects that were induced in the assembly process itself but not detected until test. Yield is central to all aspects of managing semiconductor throughput, output, delivery, quality, cost, etc. Systems must be capable of storing planned and actual yield and defect information and also the process, equipment, operator, material, etc. data associated with the production of a lot.

8.3 "Build to Forecast" for Outsourced Manufacturing

A major challenge facing fabless companies is the generation of viable forecasts, and the distribution of the build activities from such forecasts to the available capacity of the fab production and other vendors. Since these production instructions must be generated in advance of orders, the forecast demand plan is the primary input to the planning process.

Organizations typically have a number of "slices" of the demand forecast picture, and the demand planning process includes an aggregation of the slices into a combined plan. While product, or ordering part number (OPN) is generally the most important forecast key, examples of other dimensions used for the slicing include OPN group type, customer (if known), channel and manufacturing part number. A larger organization may also have demand plans sliced by division, business unit or sales region.

Within a demand plan, there are typically several levels of priority given to the demand, with current backlog given highest priority, and lower priority given to vendor managed inventory (VMI) agreements, stretch goals, etc.

Because the forecast, or demand plan, will constitute a financial agreement with the fab, it typically goes through several levels of review and approval. The approver may be working on a particular slice, or be reviewing a roll-up to overall totals.

Other checks that are typically applied to the planning process by demand planning software to automate the most common steps and reduce errors, include: carryover of unmet demand from one planning interval (typically a week) to the next; adjustment of forecast numbers week-by-week such that backlog is met (and netting is positive) by lowering forecasts where backlog is exceeded and raising forecasts where shortfalls exist, such that the total for the plan is not changed. This can be thought of as "shaping the demand profile over time," since the totals stay the same but the peaks and valleys are changed.

After the combined demand plan is approved, the next step is typically to convert the plan from OPNs to manufacturing part numbers. This will include a BOM expansion, and will identify the common manufacturing parts.

The manufacturing plan is then assigned to the outsourced manufacturing partners by generating "slices" of the manufacturing plan which total up to the capability limits of the manufacturing partners, working from the slices with the highest priority first.

If there is shared fab and assembly capacity, then the demand planning system must deal with combinations of the slices.

8.4 ERP System Solutions

In simplest terms, enterprise systems use database technology and a single interface to control all the information related to a company's business including customer, product, employee and financial data.

ERP is the current generation of systems that began as MRP and migrated to MRP II. These systems are widely used in discrete manufacturing companies around the world. The term "enterprise resources planning" was coined to reflect the fact that these computerized systems have evolved well beyond their origins as inventory transaction and cost accounting systems. The software has become the means to support and speed the entire order-fulfillment process and to automate and integrate both business and production process management.

By recording transactions, that is the computerized record of events such as the receipt of inventory or the issue of a work order, the ERP system tracks resources (e.g. materials and labor) used in financial, manufacturing and distribution management.

Today's ERP systems come in all sizes and shapes, ranging from broad functionality, highly integrated systems termed Tier 1s, to those with less functionality and/or integration termed Tier 2s and Tier 3s. Tier 1 ERP systems usually include support for:

- Finance (e.g. A/R, A/P, G/L, P/R and F/A)
- Order entry, order management and invoicing
- Purchasing and inventory
- Project costing
- Discrete manufacturing (e.g. inventory, BOM, WIP, material planning, resource scheduling and resource management)
- Human resources
- CRM

The ERP systems' planning methodology uses material requirements planning (MRP) and master production schedule (MPS) to calculate requirements for materials, make recommendations to release replenishment orders and reschedule open orders when due dates and need dates are not in phase. Many of today's ERP systems also take into consideration capacity constraints when planning production. But, unless equipped with advanced planning functionality, they do so only serially. ERP is increasingly seen as a transaction backbone and data source for ancillary, decision-support systems.

In addition, functionality has recently been introduced to support the specific needs of vertical industry segments, such as consumer packaged goods or automotive manufacturers, as well as special operations such as demand management, an essential feature for better management of supply chains. ERP systems have begun to incorporate functionality for customer interaction and managing relationships with suppliers and vendors, making the system more outward looking. Vendors are working hard to make ERP more palatable for small- to mid-sized manufacturers, particularly in the area of implementation, which can cost several times the amount of the software licenses. Other value-added aspects of the newest systems include product configuration, electronic data interchange, field service modules and Internet capabilities that extend system access to employees, customers and suppliers.

Along with functionality for enterprise and SCM, ERP also is associated with the use of client/server, relational database technology, and UNIX, Windows NT, or AS/400 operating systems. Finally, ERP can be the means for business-process reengineering, increasing flexibility and responsiveness by breaking down barriers between functional departments and reducing duplication of efforts.

As mentioned earlier, ERP systems grew out of MRP and MRP II systems, which were designed for discrete manufacturing. Examples of discrete manufacturing are automobiles and electronic equipment. The end product is an item that may be made up of thousands of components with multiple levels of assemblies.

Semiconductors are produced in a process-oriented, rather than discrete, fashion and have what is termed an "IBOM." Discrete manufacturing has a many-to-one BOM, while semiconductors have a one-to-many BOM. A particular starting wafer can produce many different types of IC chips. Each type of chip may be assembled into multiple package types. Each chip/package combination can be tested into multiple finished products. This IBOM structure creates significant problems for software that is designed for a "normal" BOM. Typical problems are tracking and storing semiconductor lot history/genealogy and product costing. To address this mismatch, many semiconductor companies use a semiconductor-oriented MES in place of the manufacturing module of their ERP system.

8.5 The Information Ecosystem: Where Communication is Key

The traditional model of an ERP system tends to be single-organization based. While it may support facilities for multiple business units within an organization, it has a hierarchical chart of business units and sub-organizations. Hence, the ERP system is at the center of the organization.

Transfers of information into and out of an ERP system are critical to use the ERP system as the basis for an outsourced manufacturing model. There are a number of technologies typically used:

- *Database-content transfers*: These follow the structure of the database itself.
- *Message-based transfers*: These are record-based feeds, similar to the real-time updates that are used in financial trading operations.
- *Ad-hoc uploads and downloads*: These are typically files that are moved from one organization to another with a file transfer protocol, then parsed.

There are standard and emerging technologies for each of these:

- Database-content transfers follow the SQL schema. Emerging technologies include XMI (XML Metadata Interchange).
- Message-based transfers include Web Services.
  - These are based on XML documents being exchanged.

  - XML documents are described and standardized through XML Schema definitions.
  - Industry forums have then defined industry-specific standard XML documents:
    - *RosettaNet* is the best example of the forum for e-commerce and manufacturing.
    - *Oasys* is the best example of a "content librarian organization" for a range of verticals.
- Ad-hoc uploads and downloads are typically Excel files and text files:
  - Excel provides a standard for organizing rows and columns, but little beyond that. Hence, data transfer management software for fabless operations management typically applies templates that aid in interpreting the rows and columns based on headers, then resolves the names and strings based on database content, constantly checking for errors and screening the content.
  - Comma separated value files are more efficient than Excel, but provide even less standardized content or structure.

Given that the data transfer challenges have been addressed, the next challenge is again one of aggregation to generate a holistic view of partner and resource status. Here, many of the same "slices" that were described above for the demand plan apply, with dimensions for the external provider organization taking the place of the dimensions for the ordering customer. Visibility could be generated via a roll-up of manufacturing status by OPN, product group, provider, business unit, etc.

The next step in the evolution of Operations Management (OM) software beyond the typical ERP solution is that once these visibility results have been generated, can they be used for valuable collaboration? The answer is, "Yes." The aggregated information is then sliced back apart according to the data security and visibility rules established for collaboration. For instance, an organization could share a demand plan with a provider organization, as long as no information specific to other providers was present in the plan.

Other types of collaboration include generation of feedback or supplier ratings, combined with listings of available supply and demand. At this point, information is not typically owned by one organization, but becomes located in a peer-to-peer "ecosystem" or information utility.

The concept of such an ecosystem or information utility is just starting to emerge, but it will be a valuable asset to the operational excellence of all the participating organizations. The variations in currency, weather changes, RoHS (restrictions of hazardous substances, known as environment compliance) all can be provided using web services SOA (service-oriented architecture) Technology.

Serus is one of the first companies that has attempted to provide services in this arena. Serus has provided many fabless companies with the ability to view consolidated WIP inventory views from snapshots and transactions that emerge from trading partners' demand from their customers' production levels.

8.5.1 Fabless Supply Chain Management

There is a class of software that is beginning to emerge in OM that is also referred to as "Fabless SCM" software. Capabilities vary from package to package, but packages contain many or most of the following capabilities:

- Fabless lot genealogy: supporting splits, combines, bonus, scrap, etc.
- WIP tracking
- WIP flush
- Wafer to die conversion
- Yield and scrap accountability
- Probe and test binout support
- Target and Actual Yield and Cycle Time storage and use in calculations
- Time phases supply/demand match
- Forecasting
- Order management
- Engineering data storage
- Available to Promise/Capable to Promise
- On-time delivery management

The packages must be capable of fetching data from the supply chain and mapping it into a set of common elements for use by the software. Vendors in this space are Serus and Oracle OSFM w/advanced planning/scheduling (APS).

Some of the challenges in applying OM are the communication with the partner organizations. There are additional challenges within the SCM module itself: partial or intelligent re-planning.

The need for such becomes clearer when the following scenario is considered. In the typical single-organization centric model, a schedule could be established in which all data is collected from 6pm to 8pm, all data is processed and aggregated from 8pm to 10pm, and the planning engine carries out a complete plan generation from 10pm to 1am, followed by a data distribution phase from 1am to 3am. However, suppose that multiple revisions of the demand plan occur at different hours of the day because the demand planners are anywhere in the world, and supplier status reports arrive at any time, because the suppliers are in time zones all around the world. There is no time window during which all content can be locked down for use by the planning engine.

Instead, a partial planning system or an intelligent re-planning system is constantly adjusting the parts of the plan that have been altered by the most newly received information, such as a change to a constraint or a demand. These partial plans should take far less time to carry out. Deeper planning occurs in background processes that are also on-going, with the deep planners using the results of the partial planner, and also then posting their results to the shared state.

This model of shifting from the fixed-schedule planning system is another foundation of the comprehensive modular intelligent OM system: in the new paradigm, all of the modules can operate in a shared continuous mode, and send partial results (which can be either successful re-plans or indicators of exceptions and infeasible conditions that need the attention of the exception reviewer) to each other through a common data model.

A diagram of the information flows between the modules of an OM system is shown in Figure 8.1.

This paradigm allows for supporting operational activities that are highly parallel and time-zone distributed. It allows for multiple levels of review of plans, alerts, and exceptions. It also allows for the generation of new plans to handle newly apparent conditions such as transportation delays, line outages, demand spikes, etc. The operations manager is then given a set of options to handle the conditions, or can generate what-ifs in which such conditions are hypothesized and tested.

8.5.2 Semiconductor Yield Management Systems

Yield management (along with circuit design) is one of the most semiconductor industry specific categories of software. The tools are typically capable of showing a map of all die on a wafer with various attributes, trends and relationships, for instance. These tools are capable of analyzing technical and operational data from throughout the supply chain in order to predict and improve yields. Like APS/SCM software, vendors focus in various areas of yield management, with no vendor covering all aspects. The main vendors in this space are ACME (later part of KLA-Tencor), Heuristics Physics Labs, IDS Software Solutions (now part of PDF Solutions), KLA-Tencor, Knights Technology, PDF Solutions, SiVerion, Syntricity, Yield Dynamics and Yield Power.

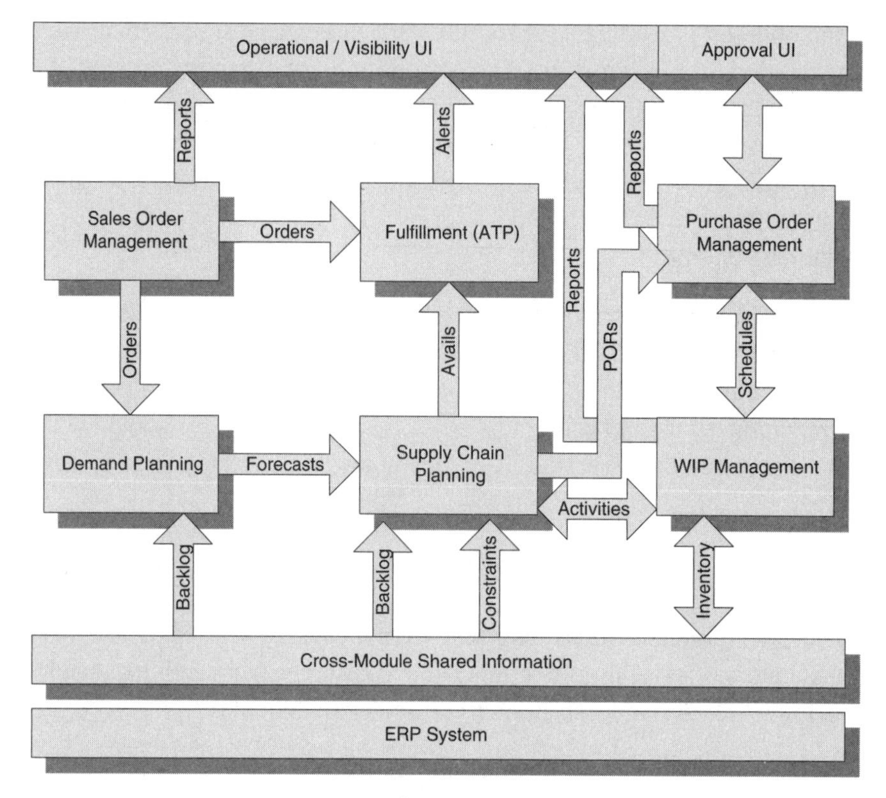

Figure 8.1

8.5.3 Product Data Management

PDM systems focus on maximizing the time-to-market benefits of concurrent engineering while maintaining control of data and distributing it automatically to the people who need it – when they need it. The way PDM systems cope with this challenge is that master data is held only once in a secure "vault" where its integrity can be assured and all changes to it monitored, controlled and recorded.

Duplicate reference copies of the master data, on the other hand, can be distributed freely, to users in various departments for design, analysis and approval. The new data is then released back into the vault. When a "change" is made to data, a modified copy of the data, signed and dated, is stored in the vault alongside the old data, which remains in its original form as permanent record.

Functionalities that may be included in a PDM system are:

- *Work Management*: manages what happens when someone works on the data
- *Workflow Management*: manages the flow of work between people
- *Work History Management*: manages all the events and movements of the two above.

8.5.4 Customer Relationship Management

CRM is an information industry term for methodologies, software, and, usually, Internet capabilities that help an enterprise manage customer relationships in an organized way. For example, an enterprise might build a database about its customers that described relationships in sufficient detail so that management, salespeople, people providing service, and perhaps the customer directly could access information, match customer needs with product plans and offerings, remind customers of service requirements, know what other products a customer had purchased, and so forth.

According to one industry view, CRM consists of:

- Helping an enterprise to enable its marketing departments to identify and target their best customers, manage marketing campaigns with clear goals and objectives, and generate quality leads for the sales team.
- Assisting the organization to improve telesales, account, and sales management by optimizing information shared by multiple employees, and streamlining existing processes (e.g. taking orders using mobile devices).
- Allowing the formation of individualized relationships with customers, with the aim of improving customer satisfaction and maximizing profits; identifying the most profitable customers and providing them the highest level of service.
- Providing employees with the information and processes necessary to know their customers, understand their needs, and effectively build relationships between the company, its customer base, and distribution partners.

8.5.5 Sales Force Automation

SFA is "the application of information systems technologies to sales and marketing activities." The sales force itself is not only being automated, but also the entire sales process. Sales is the organization which, by definition, works closely with the customers of a company. Sales people represent the company and develop relationships with customers on behalf of the company. A firm could have one of the best products on the market, but without a competent sales force be doomed to failure.

SFA is important, because it improves the effectiveness and efficiency of the sales group, which translates into greater revenue for companies. The SFA packages can perform a number of tasks including lead management, consolidation of all customer contact information, provide a repository for all sales and marketing materials, provide up-to-date product specifications, availability and pricing information, provide information regarding order status and provide visibility to customer service issues. The SFA packages can be used to aggregate information on every conversation or contact with a customer. For example, a salesperson that was planning a customer visit could access information regarding past purchases, current orders in progress, customer service issues and review summaries of all conversations with customer personnel. Also, they could review competitive analysis to understand how the company's products compete and prepare a customized presentation based on templates in the SFA system. One of the keys to a successful sales transaction is to understand a customer's needs and SFA tools help sales people accomplish this objective. Depending on vendors and views, there may be a great deal of overlap between what gets classed under SFA and CRM.

8.5.6 Manufacturing Execution Systems

MES bring pervasive computerization to plant floors in a systematic way, by placing on a single platform such diverse functions as quality control, document management and plant-floor dispatching. At its core, MES track WIP through detailed product routing and tracking, labor reporting, resource and rework management, production measurement and data collection. By capturing "live" information about setups, run times, throughput, and yields, managers measure constraints, identify bottlenecks and get a better understanding of capacity.

When the MES concept was first introduced, it was hoped that generic functionality could be defined and made applicable to any plant floor. But time and experience have led the most successful vendors to pursue a "narrow-and-deep" strategy; devoting their software development to the industries they know best. For manufacturers, the availability of more functionally robust systems leads to less software customization and quicker implementations.

There are more than 100 MES software vendors in the United States alone. Five of these have focused on the "narrow-and-deep" strategy for semiconductor process flow support. These vendors include Camstar, Consilium, FASTech, Promis and IBM with SI View.

Typical functionality provided by semiconductor MES systems:

- WIP tracking
- Process/flow/recipe management
- Engineering data collection
- Statistical process control
- Equipment maintenance management
- Specification management

- Shop-floor dispatching
- Product cost accounting

These functions are typically tightly integrated within the MES system, so that information is entered once, either automatically or manually, and is then usable for a number of tasks. The packages mentioned are specifically tailored for running semiconductor process flow manufacturing including wafer fab, probe, assembly and test. Not only can MES run internal operations, they can also be used to track, at various levels of detail, outsourced manufacturing operations.

Semiconductor MES packages are poorly suited, however, for supporting discrete manufacturing such as board assembly production, either internal or outsourced.

8.5.7 Advanced Planning/Scheduling and Supply Chain Management

A variety of software packages and software functionality gets lumped under the headings of APS & SCM. Some of the packages are stand-alone, working in concert with other applications such as ERP or MES. Other packages are "modules" offered by ERP and MES suppliers. The APS/SCM categories can be viewed from several perspectives.

The first perspective is level within the supply chain. Simplistically, partitioning the levels (with examples) as follows reveals:

- *Shop Floor*: short order scheduling and real-time dispatching
- *Factory*: shift/day/week/month starts, capacity management, order fulfillment
- *Enterprise*: multiple factory supply/demand balancing.

The second perspective is supply versus demand. Some software focuses on analyzing and managing the customer or demand side. Other software focuses on analyzing and managing the supply side. Other software works at optimizing supply/demand balance. Within supply/demand, there is an additional perspective of constrained versus unconstrained analysis.

A very popular subset of supply/demand optimization today is ATP. In its highest form, the tool is capable of providing a response in real time to a request for availability of some quantity of a particular product on a particular date. To accomplish this, the tool must have detailed visibility into the inventories and capacities throughout the entire supply chain. The typical issue here is not the development of the algorithms to provide the answer, but, rather, to have access in real time to all the data required to perform the calculation.

The third perspective is modeling tools versus management tools. Modeling tools tend to be used occasionally to determine the optimum way to configure a manufacturing operation while management tools are used on a continuing basis to manage on-going operations.

A fourth perspective is usability of any particular tool for semiconductor type manufacturing. Just as the manufacturing modules of ERP systems have limited usability for semiconductor process flow manufacturing, so do some of the commercially available APS/SCM tools.

There are no "universal" APS/SCM tools that do everything. As a matter of fact, the reality is that there are tools that are designed for a particular intersection of the various perspectives. Generally, there are no "universal" suppliers that offers tools (integrated or not) that fulfill all the various functions. Some vendors (e.g. i2 and AutoSimulations) have partnered to provide wider coverage to the various requirements.

The major vendors that support the semiconductor industry are Adexa (formerly Paragon), Aspen/Chesapeake, AutoSimulations, i2 Technologies, Manugistics and Oracle APS.

CHAPTER 9

Quality and Reliability

9.1 General

9.1.1 Introduction

This chapter includes details on quality and reliability (Q&R) management in the fabless environment. The lack of direct control of supply chain resources by fabless companies requires that they develop a sense of trust in their suppliers while keeping sufficient controls in place to ensure Q&R. The first section addresses some of the general concepts as they apply to the fabless environment. The second section focuses on front-end Q&R, which is defined here as activities and systems intended to ensure and improve Q&R of the semiconductor products. The third section tackles back-end Q&R, which is defined here as those activities and systems which deal with Q&R issues after products have been shipped to the customer. The final section discusses environment, health and safety (EHS) concepts. This section is included in the Q&R chapter because there is a significant overlap between concepts and systems in the Q&R environment and those in EHS.

Semiconductor devices enable all electronic products and many other types of products used on a daily basis, such as phones, computers, toys and cars. For most people, the semiconductor industry and its products are hidden from view, with only a few exceptions (most notably "Intel Inside" and ATI and Nvidia graphic chip logos). Each and every year, semiconductor devices become smaller, cheaper, use less energy and offer increased features and computing power. The industry is characterized by rapid new product introductions and very high expectations on Q&R. A missed product cycle or a serious Q&R issue often results in severe business consequences.

Many of today's most innovative products stem from fabless companies. Even though these organizations do not have direct control over their supply chain, they are expected to deliver products that meet the same Q&R standards as companies that own and manage their own fabrication facilities (integrated device manufacturers (IDMs)). Wafer fabs, assembly, test and logistics facilities are often thousands of miles and many time zones away from the fabless company.

New fabless companies will find it hard to sell any product without meeting some of the basic customer expectations on quality controls, which will include at least ISO9000, a well-defined supplier management approach and a basic EHS system.

9.1.2 Organizational Checks and Balances

To efficiently and effectively perform their jobs, employees must have a clear understanding of their roles and responsibilities within their organization. They should fully understand the existing documented quality management system (QMS) and how their efforts contribute to the achievement of the goals and quality objectives. The responsibility and authority of personnel who manage, perform and verify work affecting quality must be clearly defined. In particular, production and production control operations, even in the fabless environment, must identify personnel who have responsibility for ensuring product quality. Job descriptions that define each employee's general job requirements are an essential part of this approach.

Management review is a tool that is used to promote continual improvement. Management conducts periodic reviews of the effectiveness of the entire QMS and changes that could affect it. These reviews include monitoring trends in operational, business and quality performance of the product development processes and the associated support processes.

Metrics are defined for key performance areas and are used to monitor ongoing progress towards quality objectives, to identify critical issues, to track improvement activities, to measure cost of poor quality, and to identify and prioritize opportunities for quality and productivity improvements. Data and information from all sources of product and process problems, including analysis of field failures, is also reviewed to identify areas where action may need to be taken to reduce or eliminate non-conforming product and to prevent problems from re-occurring.

The goals and objectives of the organization are analyzed to ensure that they align with known customer requirements. Customer satisfaction measures, such as customer surveys and customer scorecards, are used to ensure that improvements in internal performance measurements of the product development process are resulting in increased customer satisfaction and that the results are relevant and accurate for drawing conclusions. Results of internal assessment and the status of corrective and preventative actions are reviewed, including results of follow-up actions from previous management reviews. Competitive comparisons and benchmarks are used, when appropriate, to drive improvement activities. Records of these reviews are maintained.

In larger companies, the quality role is often separated from other operational roles for obvious reasons. The quality organization is expected to represent the customers' interests, and this can only be achieved if the organization does not report into an operational group. True checks and balances can only be achieved if the highest manager of the quality organization has the same hierarchical status as the highest operations manager and the highest product development manager.

9.1.3 Quality Standards

Doing business in the electronics industry has become virtually impossible without at least ISO9001 certification. For smaller fabless companies this can be a difficult hurdle to overcome. ISO9001 requires extensive documentation of the development and manufacturing processes inside the company, which is hard to achieve if a company has very few dedicated quality resources and an overextended and sometimes skeptical engineering staff.

Fabless companies can and should leverage ISO9001 and other certifications by their suppliers. Most large wafer fabs, assembly and test providers are ISO9001 and TS16949 (a quality standard defined by the automotive industry) certified. For fabless companies this provides a basic assurance that the manufacturing processes at the subcontractors are documented and that procedures exist for change notification and for dealing with excursions.

Many resources are available to the fabless company to achieve ISO certification. Fabless start-ups should initiate the certification activities as early as possible. Building a development and manufacturing process around a set of specifications is easier than adding these specifications afterwards. Having an experienced internal auditor is an absolute requirement for a successful ISO9001 implementation. Many experts are available who can perform this type of service, but it is essential to find one with the right type of expertise. Someone who has only worked with large IDMs will find it hard to guide a small fabless start-up through the certification process.

9.1.4 Building in Quality

Most companies have realized that excellent product Q&R is not a luxury, and certainly not an afterthought. Very early on in the product definition stage consideration should be given to how the product will be tested. Testability considerations (called DFT, or design for test) have a huge impact on product architecture. Early on in the semiconductor industry, quality was ensured (at least to some extent) by extensive testing and screening after the production cycle was finished. This is no longer the case, and the impact of the Japanese quality movement and approach has been as big in the semiconductor industry as it has been in for example the automotive industry. Since the cost of correcting quality problems afterwards is huge, fabless companies (and IDMs) are pushing quality considerations as far "upstream" as possible into the product development process.

Failure Modes and Effects Analysis (FMEA) is a methodology that, if implemented correctly, will highlight potential weaknesses in the design and manufacturing flow of semiconductor processes early in the development cycle where it is easier to take actions to overcome these issues, thereby enhancing reliability and also lowering cost through design. FMEA is used to identify potential failure modes, determine their effect on the operation of the product and identify actions to mitigate the failures. A crucial step is anticipating what might go wrong with a product. While anticipating every failure mode is not possible, the development team should formulate an extensive list of potential failure modes – as extensive as possible.

The early and consistent use of FMEAs in the design process allows the engineer to design out potential failures and produce reliable, safe, and customer pleasing products. The FMEA approach also captures historical information for use in future product improvement. FMEAs should always be done whenever failures would mean potential harm or injury to the user of the end item being designed. Historically, engineers have done a good job of evaluating the form and function of products and processes in the design phase. They have not always done so well at designing in reliability and quality.

Unfortunately not all failures can be predicted. Therefore, monitoring of manufacturing process flows will always be necessary. By performing in-line monitoring, issues should be highlighted soon after they occur, not weeks or months later. There is usually a variety of inspection and test points within each process to verify that the process is in control, that product, process and customer requirements are being met, and to provide feedback for continual improvement. Inspections of final packaged products (product audits) should still be conducted, even though ideally no issues should be able to escape the in-line inspections and up-front FMEA-driven measures.

9.1.5 Metrics

To measure progress, a limited number of metrics should be chosen and monitored on a regular basis. Some companies measure only a handful of important items, while others go overboard and measure hundreds of parameters to decide what works and what does not. Picking the right metrics is key to improving quality and to decide where to deploy resources. Improving customer satisfaction is the ultimate goal of all quality improvement efforts, so whatever metrics are chosen, they should reflect the customer experience as closely and as accurately as possible, without a long lag time. Metrics reviews should be held on a regular basis, and should also be part of the management review. Some examples of groups of subjects for metrics reviews are information on audit results; customer feedback; process performance and product conformity; status of preventive and corrective actions; follow-up of prior actions; recommendations for improvement and an analysis of actual and potential field failures including their impact on quality, safety or the environment. An example of a very specific list of topics for a monthly metrics review is shown below:

- Delivery schedule performance
- Manufacturing cycle time

- Customer notifications related to quality or delivery issues
- Customer disruptions including field returns
- Customer complaints
- Customer visits reports
- Customer scorecards
- Customer hotline data
- CAR (Corrective Action Reports)
- Supplier audits
- Distributor audits
- Web surveys
- Outgoing inspection results
- RMA (Returned Materials Authorization) process performance (cycle time and failures found).

Companies typically monitor outgoing quality ("parts-per-million" PPM rate), but also try to understand the failure rate the customer actually experiences. These two can be significantly different. Customers usually do not use all features of a product, nor do they report all failures back to the suppliers. Therefore, a company's outgoing PPM rate can be significantly worse than what is reported through customer feedback. A difference of an order of magnitude is not uncommon.

9.2 Front-End

9.2.1 Reliability Testing

Q&R should not be "inspected into" a product, as discussed earlier. A product is either well designed and manufactured, and therefore has great Q&R, or it does not. Attempting to mask issues through extensive testing, stressing and datasheet errata is very expensive at best. A systematic approach that emphasizes quality at every phase of product development through manufacturing is essential. From initial design conception to fabrication, test and assembly; quality must be built-in and assured through stringent statistical process control (SPC) monitoring of fabrication and assembly processes, materials inspections, wafer level reliability (WLR), new product qualifications, reliability monitoring of finished product and strict change control management.

Reliability is defined as the probability that the product will perform its intended function for a specific period of time under defined usage conditions.

There are two basic types of failures, Early Failures and Wear Out Failures. These are reflected in the curve shown below in Figure 9.1, which is known as the Bathtub Curve. Typically early failures can be seen within weeks, days or even hours. Wear Out Failures should not occur within the guaranteed lifetime of the semiconductor product, typically at least 10 years.

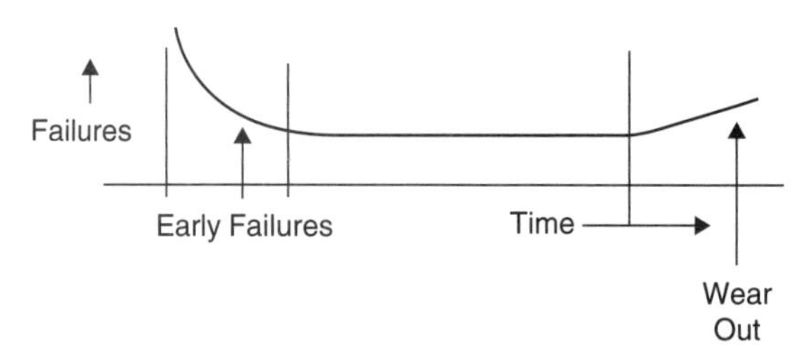

Figure 9.1 Failure Types

Reliability testing is used to ensure all products perform better than targets set for early failure rates and wear out failures. Reliability testing is performed on new wafer fabrication processes and new packages, but also after a significant change in an existing process. Typical tests include accelerated early failure testing, operating life test, temperature and humidity biased test, temperature cycling, auto-clave, ESD/latch-up and board-level temperature cycling (for packages). Power cycling and data retention testing is also performed when applicable.

Some of these tests, such as operating life tests can take up to 1,000 hours. Most companies allow limited product shipments before this test is completed. For example if no failures are observed after 168 hours (1 week) of elevated temperature testing, low-volume initial shipments can be allowed to a limited number of customers. Usually these customers are notified that not all tests have been finalized yet.

9.2.2 Maverick Controls

SPC has been used in the industry for many years, and much has been written about the subject. One of the aspects of SPC, Maverick Control, has had a significant impact on quality in recent years. While typical SPC looks at how a manufacturing process behaves, and tries to prevent and correct drift of manufacturing parameters, Maverick Control attempts of explain what the potentially negative characteristics could be of a specific production lot that is within specification, but yet different from other production lots.

Even in a very stable production environment, some lots will show characteristics that are significantly different from lots manufactured shortly before and after. Even though the lot under question is well within specification, it is measurably different. Traditionally, this material would be shipped to the customer – it is after all within specification. Seemingly random process variations occur all the time, and most of these will not impact the quality of the product. Some variations however will impact quality, and the goal of Maverick Control is to catch and isolate these lots before they create problems, and to understand what is different about this material. An example is shown below in Figure 9.2. Typically processes are controlled within the lower and upper specification limits (LSL and USL), and variations within these limits are used only to identify and if necessary correct trends. Maverick Control addresses all variations that are outside the MCL limits, as indicated below.

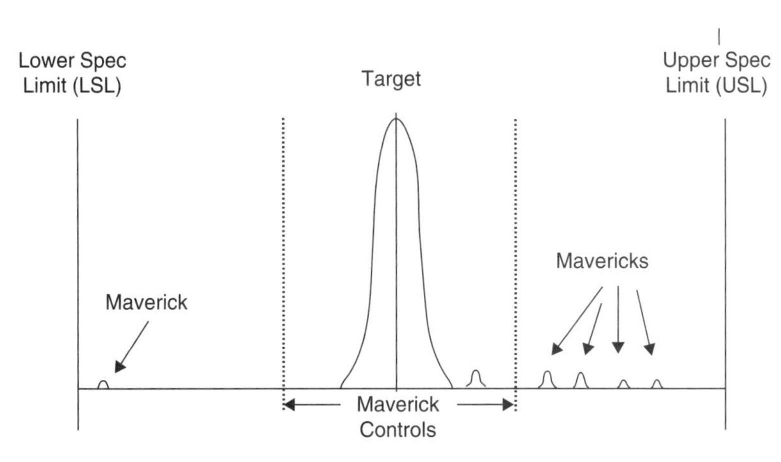

Figure 9.2 Maverick Controls

Maverick Control is a requirement for some (e.g. automotive) customers. All these efforts are driven by the fact that problem prevention is orders of magnitude cheaper than the cost of reaction. Process variations, even within the specfication limit, can be an indication of a serious underlying problem. When root cause investigation and elimination are part of Maverick Control process there will be significant improvements in quality, reliability, yield and therefore in overall product cost.

9.2.3 Reliability Monitoring

Most fabless companies (and IDMs) have a reliability monitoring program, where sample products are pulled from their finished goods inventory on a regular basis, and submitted to a subset of the stress tests mentioned above. This process will flag issues that may have developed over time in the manufacturing flow.

Results of reliability monitoring programs are usually published on the company web pages. Test frequencies vary, depending on the maturity of the fabrication process and the stage of the product in its life cycle. Reliability monitoring can be expensive, since products are pulled from finished goods to test a representative and realistic sample. Tests also tie up equipment that could otherwise be used for qualification of new products. This is a trade-off every company has to make, but some level of reliability monitoring is absolutely necessary and usually required by customers.

9.2.4 Supplier Management

Fabless companies depend on the quality of purchased materials and services. None of these activities are typically under direct control of the company, with the possible exception of product test early in the product life cycle.

Supplier management is typically based on an AVL (Approved Vendor List) procedure. Vendors (which can include wafer fabs, assembly houses, test providers, calibration labs, etc.) have to meet certain requirements to be entered onto the AVL. The goal of every company should be to keep the AVL as short as possible, because adding vendors adds cost and risk. This additional cost and risk has to be weighted against the advantage of possibly obtaining a lower price through competitive bidding. Regular audits, reviews and other mechanisms are used to maintain the AVL. Vendors who are unable to meet or continue to meet the companies' requirements can be put on probation and ultimately removed from the AVL.

A well-documented purchasing flow is essential to maintain discipline. As the number of products in a fabless company grows, so does the number of mask sets, process variations, exceptions, inter-company contacts, etc.

Clear agreements on pricing, available capacity, delivery schedules and change management are essential, as are governmental, safety and environmental regulations, which can impact many details of the interactions with vendors. An example is the long list of materials not allowed for packing the final product for shipment to the final customer. Many fabless companies never physically handle the products once they reach full production status, and clear and verifiable agreements with vendors covering all the necessary details are essential.

A vendor scorecard system is often used to drive performance improvement at suppliers, and to make decisions on which supplier will receive more business for future process generations.

On-time delivery is one of the critical parameters monitored by nearly all fabless companies. Orders to suppliers are placed based on many input variables, such as existing inventory, vendor cycle time and customer demand forecast. Not meeting on-time delivery commitments by the supplier is one of the variables that can quickly create major problems at any company. Monitoring of throughput time and meeting shipment commitments is usually weighted heavily in vendor scorecards.

Ultimately the fabless company remains responsible for all aspects of quality, reliability and delivery. Problems created by suppliers should be caught and corrected before they impact the end customer. If they are not, one or more of the supplier management systems are not working as they should and corrective action should be taken.

9.2.5 Document Control

Document Control is an essential part of the workflow of every company. Without documented processes, it is hard to be consistent, and it is also difficult to achieve improvements, since nobody will know the exact work flow. In smaller fabless companies, which are typically engineering driven, document control is sometimes seen as an unnecessary and bureaucratic activity. Following the ISO9001 spirit (document what you do, do what is documented – while implementing and documenting improvements along the way) is obviously impossible without a document control system. Turnover in personnel, organizational changes and other turmoil can cause companies to "lose the recipe" when workflows are not documented. Training of new personnel is also made much easier by good documentation.

Document control should cover most processes in the company, from strategic planning, product development, purchase orders and payroll, all the way to supplier management. Several document control systems are available commercially, and include essential features such as change control management.

9.3 Back-End

9.3.1 Product Change Notifications (Customer and Supplier)

Changes are part of life in the semiconductor industry. Changes can be initiated by the fabless company to improve certain aspects of the product, such as yield. Changes can also be triggered by the wafer fab or assembly house, for example due to equipment changes, expansion into a new facility, a merger or an acquisition.

Whenever change occurs, a well-defined management process has to be used to plan, qualify and implement this change. Where practical, analysis is performed on potential impact to the systems in which the product and process is ultimately used, but also on the effect of changes on product already delivered to the end customers.

A formal, documented change process is used to ensure that the appropriate validations are completed and that modifications are documented prior to implementing the change. When a product or process change requires the approval of a customer, a formal product change notification (PCN) process is used. Records are kept indicating the initiation of any change to production processes and to demonstrate conformance to these requirements.

Usually the end customer is provided with a document (PCN) in which the changes are spelled out, and which should include a schedule for availability of samples of the modified product. End customers have to be given a certain amount of time to verify the modified product, and to manage the transition. This often means the fabless company has to maintain production of the original product while ramping up the modified version. In general, change management is a difficult and potentially costly process. This, once again, emphasizes the need to "do it right the first time," and spend more time up-front during the planning and development stages of the product life cycle.

Some changes are initiated by suppliers. Fabless companies need to ensure that they have a well understood and documented agreement with their suppliers on what type of changes can or cannot be made without notification. Small changes occur in every wafer fab on a daily basis, and very few of these will impact the fabless companies' products. Some changes however can have unexpected impact on the behavior of the products, which may be undetectable even to the fabless company, but can impact the end customer. Once again, a clear change control procedure is essential, all the way through the supply chain.

9.3.2 Customer Returns (RMA Process)

The RMA process has been mentioned several times in this chapter. This process is one of the most important feedback mechanisms a customer has access to. A solid and responsive RMA process can significantly improve the relationship with customers. A slow, bureaucratic and ineffective RMA process can do irreparable damage to a companies' reputation.

Every semiconductor company will experience product failures from time to time. The severity can vary (from an application-related issue at a single customer, to a product recall impacting all customers), but in all cases the reaction to the problem and the expediency in resolving the problem is much more important than the original issue itself. Most customers are realistic, and realize that failures, whatever the underlying root cause is, will occur from time to time. Resolving the issues quickly can greatly enhance customer relationships and improve the customer's confidence in the supplier.

RMAs can be reported through many channels. Most companies have a hotline, which should be the initial contact for RMA issues. RMAs can also come in through sales, through executives or through direct engineer to engineer contact between supplier and customer. It is essential for any company to have a solid tracking system that does not rely on e-mails or voicemails between individuals. Progress should be reported to the customer(s) on a regular (usually daily) basis. Many customers expect and demand an initial report within 48 hours, with a full Eight Disciplines (8-D) report (see next sections) in 2 weeks or less.

As soon as the failed component(s) has been received by the supplier, the clock starts ticking. Through an effective Failure Analysis (FA) flow (see next section), the root cause of the problem needs to be found as soon as possible. Depending on the root cause and the severity of the issue, the outcome can be as simple as a dedicated application note for one customer, all the way to a complete recall of all products in the field.

9.3.3 Failure Analysis

From time to time, devices will fail at the customer, either during incoming inspection, during board manufacturing or, worst of all, after the product has been deployed in an end-user system, such as a cell phone base station or an automobile.

When failures occur, companies typically enter a process called failure analysis (FA). Depending on the type of failure, the severity of the failure and the number of failures, a number of resources have to be made available to address the issue. Customers, rightfully so, expect quick answers when failures occur.

The process of FA occurs at all levels and within all business processes of any organization. Typically, a team-centered and interdisciplinary approach is practiced. If the conclusion is that the failure is real, and that more customers may be impacted, a different process (Materials Review Board (MRB), which will be discussed later in this chapter) may be triggered.

The typical FA flow includes the following broad steps: assure failure validity, fault localization and characterization, sample preparation and defect tracing and root cause determination.

FA is a critical step that strives to discover physical evidence that clearly identifies the cause of failure. This evidence is sought through the analysis on a case-by-case basis of failed ICs. Electrical and physical analysis is performed using a combination of several straightforward and sophisticated analytical measurement systems, bench top equipment and techniques. The process is successful when the location and the cause of failure are isolated on the product.

Collaboration between many engineering disciplines, such as design, product engineering and software, but also with the wafer fab or assembly house may be needed to move the analysis forward. Analysis progress, results and conclusions are communicated to internal and external contacts supporting the Containment and Corrective Action processes to identify process owners who can implement changes to limit and eliminate the cause of failure. Some examples of failure mechanisms are electrical overstress (EOS), masking and etch anomalies, mechanical damage, electrostatic damage or discharge (ESD), gate oxide ruptures and test coverage issues.

Sometimes failures are caused by application issues. In that case, close cooperation with the customer is required to resolve the issue. Nothing can be as frustrating to the customer as receiving a report that no failure has been found, and that there must be something wrong with the way the product is being used.

9.3.4 Materials Review Board

When a potential problem has been identified through a customer complaint, a reliability monitor or any other means, the company has to quickly decide on the correct resolution.

FA usually is only one of the steps in the resolution of these issues. Typically companies go through a process referred to as Eight Disciplines (8-D) process, which is managed by a cross-functional team called the MRB. The 8 steps are:

Step 1 Establish the Team

Step 2 Describe the problem

Step 3 Develop an Interim Containment Action

Step 4 Define and Verify Root Cause

Step 5 Choose and Verify Permanent Corrective Action

Step 6 Implement and Validate Permanent Corrective Action

Step 7 Prevent Recurrence

Step 8 Recognize the Team

Step 3 involves a critical and usually difficult decision. If the problem is considered serious enough, all shipments to customers may have to be put on hold. Another difficult decision is customer notification. If there is a problem, the customer will usually want to be notified quickly. The timing of this notification can be difficult. Premature notification can create issues, since there are more questions than answers early on in the investigation. Late notification can be even more negative, since the company can be perceived as attempting to hide a problem. The decision to stop shipments and notify customers of a (potential) problem is usually a very difficult one, and many companies involve upper management in the final decision.

One key aspect of managing the 8-D flow is traceability. Companies have to be able to determine the root cause of the problem. Therefore, they have to be able to trace the exact genealogy of a product, in other words, trace the wafer fab, the wafer lot, the wafer number, the assembly lot, the test program used to test the product, etc. Quick access is needed to yield results, fab equipment excursion data and other information that can lead to the real root cause. In addition to backward traceability (which gives information on how, when and where the product was manufactured), companies also need forward traceability. If, for example, a wafer lot is determined to be bad, a company has to be able to trace exactly which customers received products from that specific wafer lot.

9.3.5 Datasheets and Errata

The definition of quality is "performance according to the specification." The specification is therefore at the core of the overall quality system. In the semiconductor industry, the specification (at least to the customer) is the datasheet, and therefore great care should be taken with datasheet creation and maintenance. During the life cycle of a product, datasheets typically go through three distinct stages: Advance, Preliminary and Final or Production. The advance datasheet is usually not complete, and subject to change without customer notification. The final datasheet should change rarely, and this is probably why so many datasheets never reach the "final or production" stage, but remain stuck forever in the "preliminary" stage.

If a product is truly unique, and is perceived to be essential to the customer's competitive position, the customer will start using this product very early in the life cycle. First silicon may not even be out, but the customer is already doing system design and possibly a board layout based on the advance datasheet. It is not hard to imagine the issues that are created by changes, even in the "advance" datasheet stage.

Errata are typically used to modify "final or production" datasheets. Great debates sometimes take place about issuing errata on "advance" or "preliminary" datasheets, since these usually state that changes can be made without notification. The most successful companies, however, always have the customer in mind, and do not become tangled up in debates on the necessity of issuing errata on datasheets.

9.4 Environment, Health and Safety

EHS management systems are designed to ensure that environmental, safety and health issues are identified, corrected and communicated to management. The system usually includes a periodic review of legal requirements, both corporate and site audits of each manufacturing facility, and reporting mechanisms designed to ensure that management is informed of any EHS-related issues. In fabless companies, EHS systems are typically much more limited in scope than those at IDMs.

At the center of the EHS management system is the self-audit program. This program includes an audit of all portions of the EHS management system, an evaluation of the site's compliance status, and reporting mechanisms designed to ensure that management is informed of any EHS-related issues. In performing the site self-audit, EHS professionals (sometimes hired on a consulting basis) examine the following general components of the system:

- General EHS management requirements
- Emergency preparedness and response programs
- Controls designed to address the facility's impact or potential impact on the environment and on the health and safety of employees
- Monitoring of EHS performance indicators
- Compliance with legal requirements
- Training, awareness and competence

Self-audit results are reported to the appropriate management as part of the management review. Further, the EHS manager annually reviews the following topics with upper management:

- Site self-audit results
- EHS performance metrics
- Corporate audit results
- Possible changes to the EHS management system and associated documents and procedures
- Summary of key regulatory and facility changes affecting the EHS systems
- Important awards, events and activities

Similar to ISO9001 certification, some aspects of the EHS system can be submitted for external auditor certification. The ISO14001 environmental management system, as part of the overall EHS system, is meant to develop and maintain a systematic management approach to the environmental concerns of an organization. The goal of this approach is continual improvement in environmental management. Environmental management at fabless companies is usually somewhat limited, but can cover important issues such as recycling of paper, glass and plastic and reduction of energy usage.

CHAPTER 10

Test Development

10.1 Simplifying Outsourced Test Development

Today's semiconductor manufacturers are under extreme pressure to reduce costs while decreasing time-to-market on increasingly complex products. This pressure has driven the trend to focus internal resources on core competencies and to outsource functions that can be better serviced by other specialized companies. For example, ASE offers a complete full turn-key approach to test as shown in Figure 10.1.

However, when outsourcing test program development, there is a significant risk because of the potentially complex test development process. This complexity, if inappropriately approached, can have catastrophic effects on both manufacturing costs and the product's time-to-market.

This chapter will explain the inherent complexity of outsourcing test development and provides a five-step systems approach to mitigate the potential catastrophes that lie in wait.

10.1.1 Systems Approach

A systems approach breaks down the complexity into individual decisions and highlights the impact of these individual decisions on the overall project. Outsourcing test development, if handled incorrectly, can defocus and drain internal resources and can have a negative impact on cost and time-to-market. However, efficiently managing the complexity, by using a systems approach, can become a strategic advantage and allow for lower costs and focused innovation.

10.1.2 Communications

The most important element of any system is to provide clear and consistent communication between various internal and external stakeholders. Trade-offs between revenue at risk and costs will require quick decisions, based on communications between multi-national companies and individuals with varying language skills. Weekly written report formats must be established to allow for maintenance of implementation momentum and for the outsource manager to efficiently manage the strategic direction and specific details required for results. Establishing the appropriate expectations up-front for these weekly reports assures the project

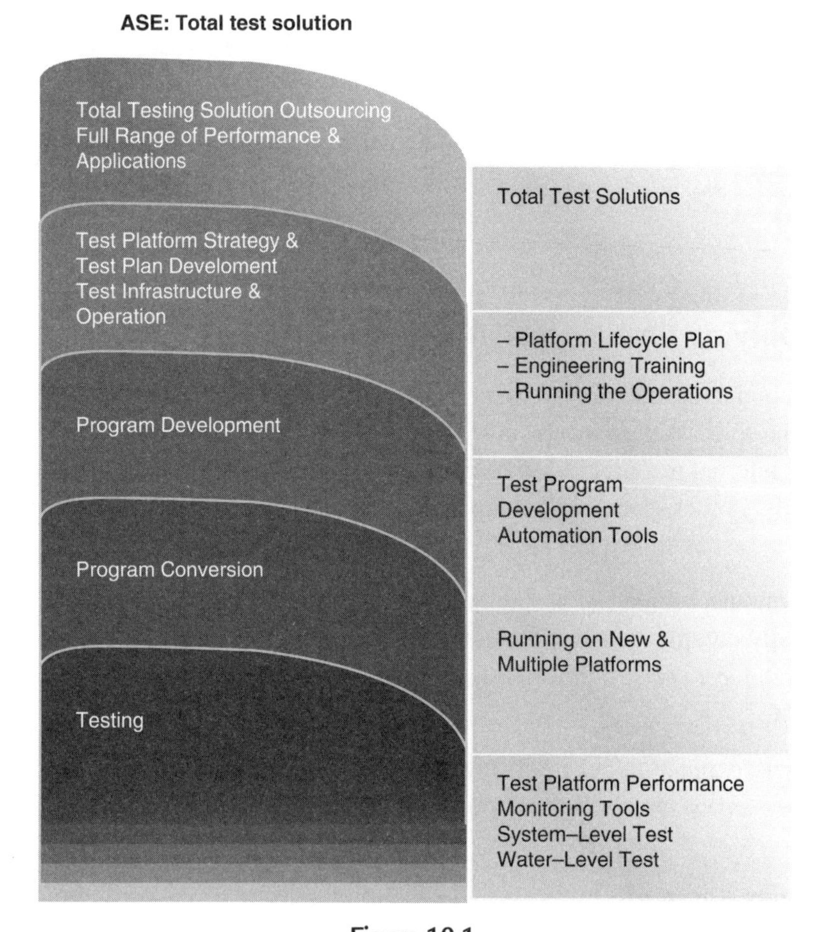

Figure 10.1

will flow efficiently and help avoid any misunderstanding during the execution phase of the project.

10.1.3 Five-Step System

The outsourcing effort can be broken down into five individual parts defined as preparation, evaluation, execution, deployment and feedback, as shown in Figure 10.2.

10.1.3.1 Preparation is the generation of an all-inclusive comprehensive description of the project. The goal is to generate a detailed request for quote (RFQ) document that contains all the information needed by a subcontractor to accurately bid the project.

10.1.3.2 Evaluation functions as a vehicle to make critical decisions. Here the analysis and feedback of the supplier's proposals takes place, the generation and acceptance of a scope of work happens, and the selection of the tester and handler platforms is defined.

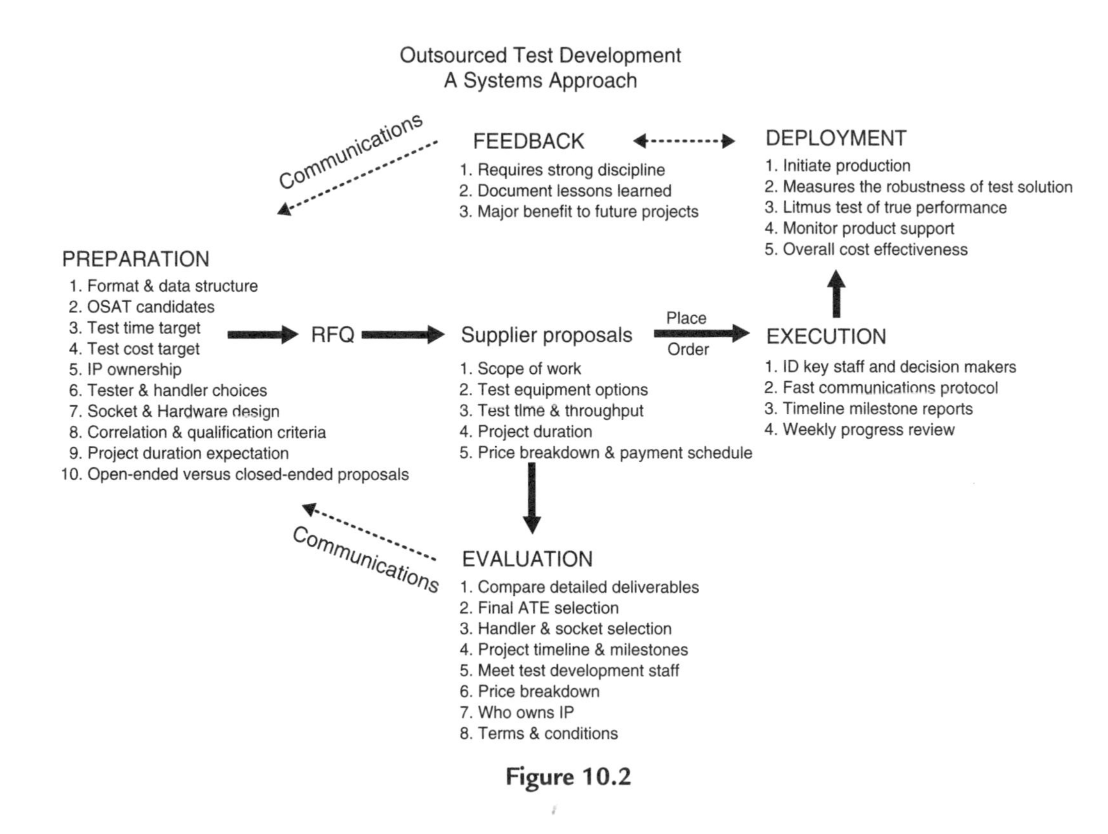

Figure 10.2

10.1.3.3 Execution monitors the supplier's service and schedules and reports are reviewed on a weekly basis. If the preparation and evaluation portions have been well executed, the execution part should flow smoothly, and unpredictable roadblocks can be quickly removed.

10.1.3.4 Deployment is the practical application of the engineering service into a manufacturing environment; it is the litmus test for the robustness of the test solution and the final exam for the development engineer's true expertise.

10.1.3.5 Feedback is an extremely important function and requires the greatest discipline. Feedback takes place throughout the project but is most important at the end of the project, in the form of a lessons-learned step. This learning curve documentation provides a return loop back to one of the stated ways of effectively dealing with complexity in future projects by leveraging experiences to simplify the complexity of the next test development endeavor.

10.2 Preparation

The resulting RFQ document must contain two key elements, which include the generation of a detailed test list and a device/product data sheet. The test list should include both characterization and expected tests that will be done in production. These are required by all test development contractors to provide an accurate quotation. The RFQ provides both the information you supply and the information you request.

It is also helpful to provide expected typical measurement data for each test in the test list, and a test methodologies document that shows specific device set-ups and idiosyncrasies that would not be apparent in the review of a device data sheet and applications hardware schematics and layouts.

The following additional information should be defined:

- *Internal Formats*: Each company should define preferred formats and data structure in advance before providing design simulations and digital pattern vectors to its outsourced development subcontractor to assure the service provider has the capability to convert them to the tester format. Other format standards should also be specified for test measurement parametric data from the test program and the software format in which hardware design schematics and layouts should be delivered.
- *Offshore Subcontract Assembly & Test Manufacturers* (*OSAT*): OSAT candidates should be selected and the chosen outsourced test development subcontractor should have an established relationship with the OSAT suppliers.
- *Test Time Target*: It is important to specify a target test time and to understand trade-offs associated with test program development. It should be determined if the test program will be modular for the shortest test time or written in a stand-alone way so each test can be removed without affecting other portions of code. Companies should decide in advance if they plan to reduce the number of tests over time via data collection, because statistical tools will be needed to evaluate this so the data collection requirements can be handled appropriately.
- *Test Cost Target*: Determining the exact cost of test is not simple so it is important to provide the supplier with a target test cost, which is typically based on the cost-per-hour of the test platform, divided by the throughput of the product-per-hour. Both of these factors can have a profound impact on the cost of the test over the long-term life of the product. At final test, it is always wise to consider taking advantage of multi-site testing, but it is important to understand how the tester resources are used versus the way the program is written related to the handler indexing parameters.
- *Ownership of Intellectual Property* (*IP*): Hardware designs and software source code, if owned, can potentially be reused or moved to other OSATs. IP ownership associated with the project needs to be defined in writing to avoid any confusion when the project is completed.

- *Tester or Handler Platforms*: Any preferred tester or handler platforms should be specified as the recommended choices. When using creativity and compromise, a significant amount of money can be saved by making the right tester selection.
- *Sockets and Hardware Design*: Provide information on socket performance requirements and, when possible, avoid using custom-made sockets. The practical life of a socket or contactor pins can have an impact on the device test cost.
- *Correlation and Qualification Criteria*: Companies must define how they will decide if the project is complete. Specifying the qualification process required and the acceptable Cp/Cpk values for each test is necessary. This will determine the robustness of each test on a specific tester platform.
- *Project Duration Expectation*: The test development process is complex, but by using a systems approach and clear communications realistic expectations and an acceptable project delivery schedule can be established up-front.
- *Open-Ended versus Closed-Ended Proposals*: This is a measurement of risk/return trade-offs between the customer and the service provider. The more risk companies are willing to take, the lower the open-ended proposal price will be. In the case of a one-price turn-key solution from the supplier, the supplier assumes a majority of the risk for any problems or delays, which will result in a higher price.

Companies must ensure the RFQ states that the supplier must provide details for the following information:

- A complete scope of the work to be performed
- The recommended equipment set for testing (tester platform, automated handler and test socket or contactor)
- An estimate of the device test time and throughput
- Estimated project duration
- Price breakdown

These items are the foundation of a basic outsource development proposal. Since this proposal will typically act as a written contract, it should include terms and conditions and a statement on IP ownership.

Multiple bids should be requested and a thorough comparison of the strengths of each proposal should be performed so they can be incorporated into the final agreement. Establishing the communications path at an early stage is important, as it allows for the outsourced parties to have access to the design and allows product engineers to have access to timely answers to technical questions. Companies should be aware they may be asked to pay for a full detailed development proposal because the supplier is being asked to potentially put several weeks worth of engineering resource time into providing a highly detailed proposal. It typically takes 2 to 3 weeks from the time the RFQ is provided to receive a detailed proposal.

10.3 Evaluation

The evaluation functions as a vehicle for making critical decisions. A breakdown of the information required in the development proposal is listed below:

- *Scope of Work*: The scope of work documentation drives the resources in the project and defines what the contractor will provide and what technical support they will require. The document should spell out what physical deliverables (work product) will be provided to measure actual performance.
- *Automatic Test Equipment (ATE) Selection*: The selection of the test platform is the most important decision a test engineer makes because there are a series of trade-offs that will have a major impact on the overall cost of test; therefore, it must be unbiased. A careful evaluation of the device specification and test list can determine the appropriate set of tester platforms. The special requirements and idiosyncrasies of the device, such as clock speed and power consumption, should be addressed in the proposal. Once the tester platform is chosen, a specific target tester configuration and minimum operating system software revision must be documented.
- *Test Platform*: Selecting the right platform will save money on development costs so it is important to carefully evaluate lower-cost testers' performance for throughput, reliability and repeatability. Determining if an upgrade for the tester computer processor equipment will impact test pricing.
- *Automatic Handler and Socket Selection*: Companies should get specific when it comes to handler and socket selection. Knowing the performance of the exact package type in a specific handler and socket is very important, as is knowing the tester configuration, and specifying a target tester that has the appropriate resources to handle the number of test sites being designed to and the handler's capability. The tester handler selection should be practical and should not adversely limit a company's manufacturing flexibility and cost.
- *Project Duration*: An accurate schedule should be required. Appropriate project management software tools can monitor the project details, including checkpoints for key decisions and handoffs. Establishing clearly assigned responsibilities for specific tasks, and incorporating the use of working groups to deal with the technical device details is vital. This will help identify bottlenecks and improve overall project performance, which will shorten the development cycle and improve the value of the

service. This will reduce time-to-market of both the overall development process and time-to-volume.

- *Test Engineer Team*: The designer should meet with the actual test engineer to make sure there is compatibility and continuity. Working with the test engineer, product engineer and process engineer in the early stages of the development process allows the team to have a positive impact on the cost of manufacturing of the product. Be wary of "store front" test development services using subcontractors or moonlighters that are not under their employment because this can create significant risk to the project completion and cause potential conflicts-of-interest.
- *Planning Durations*: Use realistic planning durations, and be aware when contractors are not providing information or being overly aggressive, especially on tester hours. Consider providing a financial incentive if a supplier completes a project ahead of schedule or a financial penalty if the project is unreasonably delayed.
- *Price Breakdown*: The proposal should list the deliverable components of the project, and each of these components should have a price. Engineering services should be separated from material costs. Don't be fooled by a lower engineering-per-hour charge; paying to train an engineer can result in very expensive debug time. A highly experienced engineer can do the work of a novice in half the time and, more importantly, will spend much less time debugging on the tester where a company might otherwise be paying hundreds of dollars per hour for inexperience. Requesting that all potential contractors provide similar project experiences and recent customer testimonials is a good rule-of-thumb.
- *Payment Schedule*: A payment schedule should be established to understand the cash flow requirements and potential financial ramifications of an exit strategy.
- *Capital Equipment*: The supplier should have contractual access to the capital equipment on which the development work will be implemented. When times get busy, capital equipment availability can become limited and the potential delays and increased rental prices can be devastating to the project schedule and budget.
- *IP*: Understand the IP associated with the application, and specify who owns the hardware and software at the end of the project. Is the program built to be stand-alone, or are there libraries and binaries that prevent portability? Request that the software code be well commented and provided in both source and binary format.

10.4 Conclusion

Semiconductor manufacturers are under extreme pressure to reduce costs and decrease time-to-market, which has driven the trend to focus internal resources on core competencies and to outsource functions that can be better serviced by other specialized companies. However, outsourcing test program development has significant risks because of the potential complexity and, if inappropriately approached, can have catastrophic effects on both manufacturing costs and the product's time-to-market.

The risks for outsourcing test program development can be significantly reduced by using a five-step systems approach that breaks down the complexity into individual decisions and highlights their respective impact on the overall project, which helps avoid the potential catastrophes that lie in wait. Implementing this systems approach can become a strategic business advantage that provides lower costs, focused innovation and can have a positive impact on the way the company chooses to compete in the marketplace.

PART 3

Becoming a Best-in-Class Fabless Company

CHAPTER 11

Best Practices for Fabless Companies

11.1 Achieving Best-in-Class Operations Practices

Having just received its first round of funding, the emerging fabless IC company has many technical, marketing and sales challenges, and operations issues are generally not on the minds of the principals. The objective of this chapter is to outline the operations processes and practices that must be addressed at emerging fabless companies to position them as a leading or best-in-class fabless company.

In established companies, the concept of "best practices" has usually been explored and implemented in some or all of these business process/practice areas. The four key steps to determining best practice are: current assessment, benchmarking, cost/benefit analysis and revising the process. Establishing best practices involves integrating business processes. It allows fabless companies to measure and improve their efficiencies, improve effectiveness of working relationships and the environment within the organization, as well as with customers and their supply-chain partners. However, establishing a best practice is also time-consuming and expensive. Judicious business decisions need to be made to elevate established practices into best practices.

Since managing cash flow is a very important aspect of life as a start-up, operations processes and practices must be in place during the various stages of development and production ramp. An understanding of these processes and practices will be of assistance to senior managers at emerging companies. Practice areas will include engineering, quality, customer support, production control and finance.

11.1.1 Operations Effort and Resources

Operations tasks and practices vary as an emerging company grows. Figure 11.1 is a conceptual illustration of the operations resources required. A model has been developed for calculating order of magnitude (OOM) estimates of operations effort and cost as a function of company revenue. For simplicity, assume a single-product company, a first-time-right design approach and three annual revenue "plateaus" – \$1 million, \$10 million and \$100 million. Outsourcing of production wafer fabrication, assembly and test is also assumed. In addition,

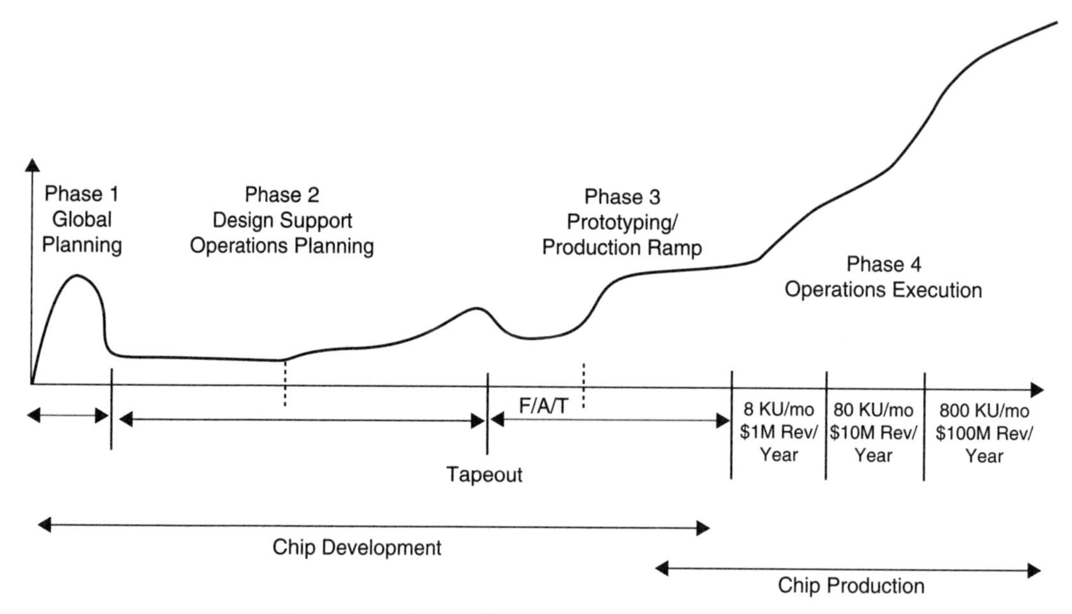

Figure 11.1 Operations Resource Requirements

assume that the fabless company will manage the supply chain and will have ownership for the work-in-process (WIP). An overview of the typical operations-related processes and practice areas that a leading fabless company needs to have in place will be presented first. The effect of using alternative sourcing methodologies will also be discussed briefly.

11.1.2 An Example on Operations Resource Requirements

As an example of some of the production management challenges at the emerging company, a summarization of the number of units shipped and the number of wafer lots that must be managed for the three revenue plateaus is shown in Figure 11.2. It also shows an estimate of the number of operations resources required to support the three revenue streams. The model also estimates the operations expenses, both manpower and other infrastructure, as a fraction

Revenue/Year	$1M	$10M	$100M
Units Shipped/Year@$10 ASP.M	0.1	1	10
Wafer Starts/Month	14	140	1400
Wafer Lot Outs/Month	0.7	7	70
Operations Resources Required	**1–3**	**3–5**	**10–17**
Ops Cost to Rev Ratio	**15%–35%**	**5%–7%**	**2%–3%**

Figure 11.2 Estimate of the Number of Operations Resources Required to Support Revenue Streams

of company revenue, which is a good measure for senior management to track their operations investment. Embedded assumptions are that the chip size is 6-millimeter × 6-millimeter, 130-nanometer 1P6M CMOS logic technology, 200-mm silicon wafers, a $10 ASP and $100K per man per year. It should be recognized that these are typical guideline numbers that could vary from company to company due to individual company specifics.

11.1.3 Operations Activities and Processes in Production

Within the fabless company, operations usually has the ownership to procure and deliver product to customers on schedule and at a favorable cost. The following is a simplified view of operations tasks and activities. Fabless operations groups should also be the focal point for coordination of cross-functional issues with customers and suppliers. Some of these roles and responsibilities can be shared with other parts of the organization. For simplicity, assume that operations is the "center of the universe" within the company. How the total company manpower can be optimized through sharing with finance, sales and engineering is beyond the scope of this article.

Operations activities are categorized into three major practice areas:

1. Manufacturing Engineering
2. Quality, Reliability, Documentation and Procedures
3. Business Processes

In an emerging company with a $1 million revenue stream, the major focus in operations should be on manufacturing engineering, while the quality and business processes can be accomplished with an ad hoc set of processes. A "systematic" focus for establishing processes and practices is recommended as the company approaches a $10 million revenue stream. A company must have established practices in all areas as its revenue stream approaches $100 million. The company may choose to invest in establishing a best practice only in the areas that are critical to its core strengths and success.

11.1.3.1 Manufacturing Engineering

This function is fulfilled by a group that usually has a broad set of technical responsibilities, such as maintaining the yield and quality of the product, interfacing with the supply chain and providing technical support to various business groups. The following is a list of typical responsibilities. Formalizing and documenting the processes associated with each of these responsibilities is the first step towards establishing a best practice in this area:

- Definition and Documentation of the Product, Procedures and Reports to Monitor and Control Product Yield, Cost and Performance
- Supplier Interface to Address Product, Design, Process and Test-Related Issues

- Yield Monitoring and Tracking:
  - Lot-to-Lot Variances
  - Wafer-to-Wafer Variances
  - Design Sensitivities
- Yield Enhancement Projects, if any
- Yield Issue(s) Resolution
- Quality and Reliability Issue(s) Resolution
- Quality and Reliability Maintenance Coordination and Execution
- Engineering Support of Customer Returns
- Failure Analysis, Debug and Reports
- Engineering Support of Production Planning
- Cost Reduction, Including Test-Time Reduction, Price Negotiations, Supplier and Technology Migrations

11.1.3.2 Quality and Reliability

This function plays a key role towards the graduation of a fledgling start-up into a best-in-class IC supplier. Setting up, monitoring and maintaining the quality processes within the fabless IC company and its supply chain are the major responsibilities. The following is a list of items around which processes can be set-up as a first step towards establishing a best practice.

Quality Manual Documentation

- Documents the Company Quality Policy, Quality System, Organizational Responsibilities and Control Mechanisms

Formalize Document Control and Related Procedures

- Verify Product Design, Manufacturing and Qualification Documentation

Product Qualification Documentation

- Qualification Plan, Specification and Report
- Supplier Data: Foundry, Assembly and Test
- Incoming and Outgoing QA Specifications
- Reliability and Quality Maintenance Program

- Process Certification Procedures and Maintenance
- Statistical Process Control (SPC)
- Yield

Continuous Improvement Program

- Certification and Reliability Maintenance
- Supplier Audits
- Change Control Procedures
- Incoming and Outgoing Quality-Level Improvement

11.1.3.3 Business Processes

These are some of the processes that seldom receive the priority they deserve. Since most companies need to establish basic financial processes early in their existence, the perception usually is that extending these processes to include operations and supply-chain issues will be straightforward. While the operations activities can and should leverage existing business infrastructure at the company, a comprehensive look at developing the proper processes required to become a leading fabless IC company should be completed. The processes are categorized into three major areas. Formalizing and documenting the processes in each of these areas is a first step towards establishing a best practice:

- Production planning and control (PP&C)
- Demand forecast
- Order management and fulfillment
- Supply-chain management
- Financial Processes
- Cash flow for WIP and inventory
- Material and cost tracking
- Product cost – actual versus plan
- Customer support
- Response
- RMA (Returned Materials Authorization)

PP&C: Demand Forecast Generation Processes

Operations plays a key role in pulling the demand forecast together. This is clearly a multi-disciplinary function that can be coordinated by different pieces of any organization – sales, marketing, finance or program management. Since operations is the group that must communicate the build forecast to suppliers and manage the order fulfillment process, they are an excellent candidate for this very critical function. Operations role involves the following:

- Coordinate cross-functional process
- Receive sales/marketing input
- Receive "do-ability" from engineering and suppliers – quantity, schedule, quality, …
- Coordinate "judged" forecast generation
- Keep updated demand forecast
- Communicate forecasts to suppliers
- Manage capacity allocation

PP&C: Order Management and Fulfillment Process

The following is a list of processes that must be followed and documented. In a best-in-class operation these activities are usually linked together in an enterprise resource planning/ material requirements planning (ERP/MRP) system connected to the company's financial system to control inventories, cost, cost of goods sold (COGS) and lead time.

- From/to sales
  - Verify availability and pricing
  - Receive/acknowledge purchase orders
- Order entry
  - Enter order into WIP/inventory management system
  - Trigger shipment from inventory, WIP or new starts
  - Notify production control
  - Acknowledge purchase order and advise delivery date to customer
- Parts shipment and logistics
  - Monitor ship readiness and advise sales 48 hours prior to shipment
  - Generate shipping/freight forwarding/import/export documents

  - Generate drop ship forms, if required
  - Pack
  - Follow through with sales and customer
- Invoice and payment
  - Generate and send invoice
  - Receive payment

PP&C: Supply-Chain Management

- Relationship management – management and working level
- Logistics management – coordination and reporting of the following:
  - Order placement, forecast communication/discussion
  - WIP, yield, quality information flow and dissemination
  - Invoicing, royalty payments (if any), cash flow management

11.1.4 Financial Processes

In executing these activities it is assumed that the operations group works in conjunction with the company's finance group but is held accountable for achieving the operations goals.

- *To/from customer(s)*
  - Purchase order receipt
  - Order fulfillment
  - Invoice issuance
  - Payment receipt
- *To/from suppliers*
  - Forecast
  - Committed forecast
  - Order placement
  - WIP ownership and liability
  - Inventory ownership and liability
  - Special materials order liability, if any

- *Management reporting*
  - COGS – plan versus actual
  - Cash flow
  - Demand versus build forecast reconciliation

11.1.5 Customer Support Processes

Having an efficient and responsive customer support organization with the associated practices is another key element of a leading fabless IC company, and these processes are categorized into three areas – quality assurance, customer service and product planning. The interface to the customer needs to be clearly identified – it could be a sales person or a program manager (Figure 11.3).

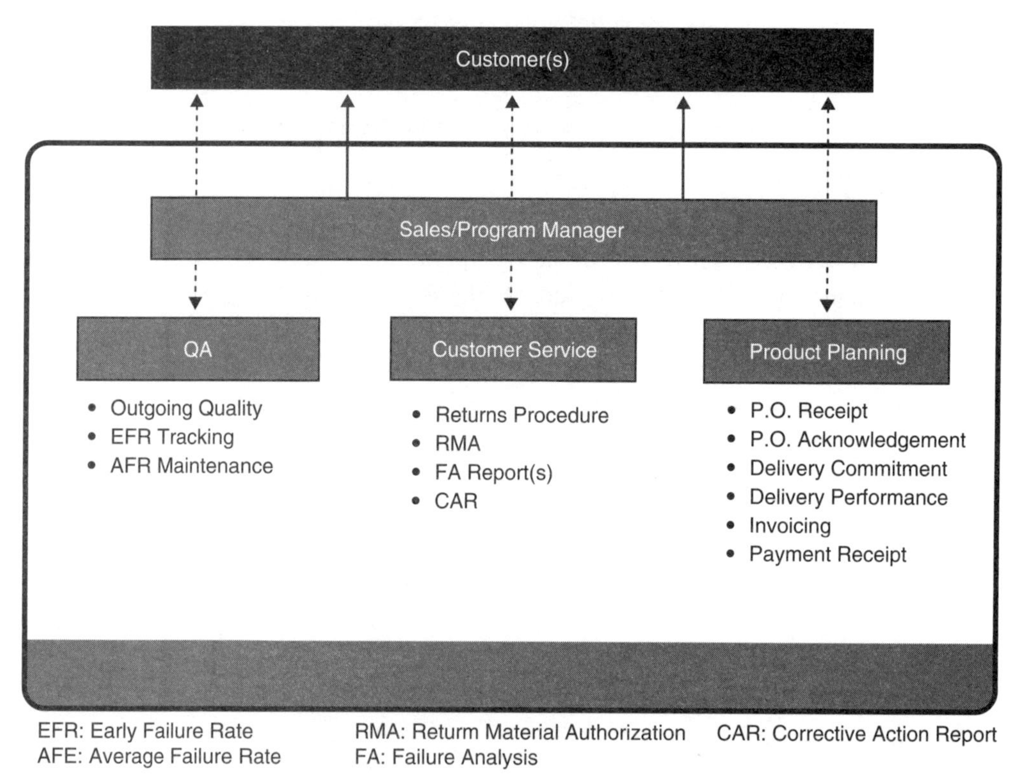

Figure 11.3 Fabless IC Company

11.1.5.1 Effect of Alternative ASIC Sourcing Models

In this particular example the customer-owned tooling (COT) methodology is considered; however, there are at least two other approaches that are important to consider – using an application-specific integrated circuit (ASIC) supplier and an ASIC aggregator. It is important to touch on this here because there is a general perception that the operations infrastructure required at the fabless company is eliminated when using these alternative methodologies. The requirements for operations best practices get reduced by 30 percent to 50 percent, but not eliminated.

Since the fabless company is responsible for product quality, delivery and order fulfillment for their customer(s), it is very important to focus on establishing all the other practices in this example. While there is some cost reduction associated with the reduced operations infrastructure, this must be traded off against the higher unit price usually charged by the ASIC supplier/aggregator.

11.1.6 Conclusions

Emerging companies aspiring to be classified as a leading IC supplier need to make an investment in establishing many best practices in the operations area. This section has provided an overview of the operations processes and best practices required as the company evolves from a start-up to a mature IC supplier. The core best practices must be in place regardless of the sourcing model used. While the ASIC sourcing model requires less resources compared with the COT approach, the fabless company must trade-off the reduced operations cost with a (generally) higher unit price.

11.2 A Foundry Manager's Role in a Fabless Company

11.2.1 Introduction

The fabless model has not only spawned dedicated foundries, but also a mission-critical organizational entity – the foundry operations manager. The foundry operations manager's contribution to the value-chain of a fabless company is extensive and overarching, resulting in improved product development and supply management efficiencies. Without a foundry manager, a fabless company lacks a core competency essential to its competitiveness. A brief description of the foundry manager's activities within the framework of an organization's value-chain is provided to clarify this role (Figure 11.4).

11.2.2 The Value-chain

A fabless company's value-chain is a collection of activities that needed to design, produce, market, deliver and support its products. Specifically, a company's value-added activities can be grouped into 10 distinct categories: design, inbound/outbound logistics, operations, marketing/sales, service, human resources, procurement, technology development and infrastructure.

Figure 11.4

The first six categories, referred to as the core group, encompass functionally critical activities and, if managed effectively, can provide a competitive edge, especially in the delivery of new and existing products. Design is included in the core group due to the fundamental value IP adds to a fabless company.

The remaining four categories comprise the support group whose activities such as training, purchasing and managing enable organization-wide execution. While the foundry manager contributes to core and support groups to a measurable degree, the major roles and activities are concentrated in the following categories.

11.2.2.1 Design

Product development is highly interactive process especially when there are super-ordinate goals of improving time-to-market and product manufacturability. As such, the design and foundry operations groups have extensive communication in all phases of a new product development cycle. Within the area of design, the foundry process and producing prototypes are pertinent responsibilities of a foundry manager.

Foundry Process

A foundry manager must be a process expert and work with the design team to specify process requirements by evaluating capabilities of current and potential foundries such as SPICE models, design rules, electrical rules, process flow and other technical specifications are reviewed. The foundry manager also directs process development and improvement activities as would be required by a new or existing product.

Prototypes

One of the most critical activities for a foundry manager is to coordinate internal and external resources to produce and deliver first silicon samples. A foundry manager must bring together suppliers and subcontractors for prototype delivery, which is essential in achieving competitive time-to-market cycle times.

11.2.2.2 Inbound/Outbound Logistics

This category involves activities for staging foundry work in progress (for on-time delivery), incoming inspection, finished goods warehousing, material handling, delivery systems, inventory control and transport scheduling. The foundry manager helps to define incoming inspection and acceptance criteria for engineering and production material. The foundry manager also specifies outgoing RMA and engineering lot handling procedures.

11.2.2.3 Operations

Operations encompasses foundry production, sort testing, final testing, quality assurance testing, assembly, equipment maintenance and facility operations – all the activities necessary to transform material inputs into a product's final form. A foundry manager assists in contracts, personnel building, controlling tapeout and mask revisions and production planning.

Contract Administrator

The foundry manager administers and helps enforce a foundry agreement. The foundry manager defines short-term goals and manages cross-functional tasks.

Relationship Builder

A foundry manager must be able to builds effective relationships with all foundry personnel, nurture a number of critical customer supplier pairs to link similar functional groups such as transportation and planning departments. Creation of these pairs greatly improves real-time execution.

Tapeout and Mask Revision Control

Tooling control is essential to the successful development of new products and manufacture of production runs. Fatal errors in the tooling database or mask-making specifications can be devastating to the development cycle time. Hence, the foundry manager participates in the database verification process before and after tapeout.

Production Planning

The foundry manager provides cycle time, capacity, cost and line yield statistics, which enables the production planners to more reliably start and stage material as well as predict delivery dates to the customer.

Electrical Test (E-Test)
The foundry manager is responsible for defining and monitoring compliance of E-test specifications.

Yield Improvement
Yields at all stages in the manufacturing process must be continually monitored and improved to reduce marginal unit costs and keep products competitive. The foundry manager facilitates yield improvement efforts by managing internal and external engineering resources.

Cycle-Time Reduction
Competition demands a sustained and managed effort to bring new products into the marketplace quickly. The foundry manager is the owner of cycle-time improvements in the supply chain, as has an understanding of both the technical and logistical aspects of IC processing.

Strategic Foundry Inspections
Facility inspections are essential in assessing how effectively manufacturing and institutional processes are being managed by the foundry to optimize product cycle time and yield. Moreover, an on-site inspection provides real-time manufacturing and quality data and statistics.

Information Storehouse
All functional entities depend upon accurate and timely subcontractor information. A large amount of non-proprietary subcontractor data is maintained by the foundry manager. Data such as supplier performance ratings are gathered, evaluated and periodically reported to internal customers.

11.2.2.4 Marketing and Sales

Marking and sales is usually described as separate and distinct from the "manufacturing chain;" however, its activities are critical and essential in an organization's value-chain. Activities such as advertising, promotions, quotations, channel selection, order processing and all aspects of the marketing-mix typify this category. Indeed, in attempting to satisfy customers, a fabless company must coordinate all its activities around this central category, including foundry management.

Feasibility Analysis
The foundry manager helps with creating and maintaining marketing mixes. When new products are defined, the technical and production capabilities of a foundry are evaluated in addition to an examination of internal resources. New market opportunities may be identified by taking advantage of superior or advanced fab processes. When existing products are

modified, the foundry manager provides essential information on whether retooling and validation activities will be cost-effective or produced in quantities required and within a time frame attractive to a targeted market.

Pricing Analysis

The consumer's perception of a successful "exchange" is increasingly focused on price, thus establishing competitive pricing policies is essential. The foundry manager provides current foundry and other applicable supplier cost data from which marginal cost and marginal revenue figures are derived. In addition, since pricing decisions are made throughout a product's life cycle, the foundry manager, with other functional groups such as finance and design, lay out a cost-reduction path.

11.2.2.5 Procurement

All the activities in obtaining purchased goods and services go through the procurement department and the foundry manager must fulfill several key roles.

Procedures and Documentation

The foundry manager ensures that procurement procedures and specifications for foundry parts are clear, accurate, consistent and compliant.

Negotiator

Price, quality, quantity and scheduling aspects are negotiated by the foundry manager. Acceptance criteria is also negotiated and specified.

Foundry Benchmarks and Statistics

Gross die-count, cycle time, capacity, line-yield, product yield and other measures of supply-chain performance are provided to help make decisions that keep deliveries on schedule and minimize order inventory excesses and shortfalls.

11.2.2.6 Technology Development

Technology development in the fabless company's value-chain category consists of any activity in creating and improving internal processes. Everyone, including the foundry manager, contributes to the on-going effort, for example, the foundry manager maintains metrics and manages change in improving product development cycles. The foundry manager identifies and creates new ways to manage foundry-related information flow between departments.

Benchmarking and Implementing Change

The foundry manager has an active role in enacting benchmarking studies of new product development processes and supply-chain flows. The foundry manager identifies and implements competitive practices and eliminates non-value-added activities.

Firm Infrastructure

This category includes activities in general management, accounting, legal, finance and strategic planning. The foundry manager is synonymous with cross-functional management and routinely coordinates activities across the core categories to improve cycle times and yields. This role constantly scans the foundry environment and plays a pivotal role contributing to the foundry selection and strategy formulation.

In conclusion, a fabless company has an intrinsic need for a multi-tasking and empowered foundry manager. The foundry manager fulfills a number of mission-critical roles such as a negotiator, liaison, team builder, problem solver, engineer, program manager and functional strategist. This role takes charge of manufacturing and delivering engineering prototypes. Moreover, the foundry manager is a cross-functional leader with activities that drive organization-wide improvements, namely, the optimization of new product development cycle times, product yield and supply-chain performance. With current knowledge of foundry and other supplier costs and capabilities, this position also contributes to the on-going evaluation of marketing strategies. In summary, the foundry operations manager significantly enhances the value-chain and competitive edge of a fabless semiconductor company.

11.3 Closing the Loop: Understanding the Manufacturing Flow

The past decade has seen a shift in the fundamental business model for bringing high-value semiconductor products to market. The increasing adoption of outsourced manufacturing, together with favorable financial markets, has led to dramatic growth in the number of fabless companies. The emergence of the fabless semiconductor market segment has largely been a result of the opportunity for rapid entry into markets where premiums are associated with being the first to market at high volume.

The fabless outsourcing business model has been most successful when new, complex products are brought to market using well-understood, standard process technologies with well-characterized IP and primitives as building blocks. The current design and manufacturing challenges of clock skew, signal integrity, static leakage current and within-die variation associated with the 130-nanometer process node will intensify as the industry drives toward the 90-nanometer process technology node and beyond. These issues in concert with the fabrication challenges posed by optical proximity correction (OPC) and other resolution enhancement techniques will continue to increase the importance of understanding parametric and design-limited yield factors.

In addition to the four key success qualities: market and customer understanding; relentless focus on execution; relentless focus on costs; and management team – given the right product with the right team, the long-term success of a fabless company will be largely determined by how well it executes in two critical areas:

- How quickly it can deliver the right product, at a profitable volume, to market, that is yield.
- How efficiently it can identify and resolve product and yield-related problems once it has delivered the product to market and is in volume manufacturing.

As designs become increasingly more complex and process technologies move to ever-smaller geometries, the importance of rapidly solving product issues cannot be overstated and will become increasingly necessary to a fabless company's continued existence. Before yield and product issues can be resolved, however, a company must develop a clear understanding of their manufacturing chain and how manufacturing variation affects their product. The successful fabless company must "close the loop" among design, product engineering and manufacturing functions. Closing the loop and efficiently solving yield and product issues require close collaboration with suppliers and continuous access to data from the manufacturing supply chain.

11.3.1 Challenges and Barriers

Maintaining access to all production data as wafers and parts move through each segment of the chain is much easier said than done, particularly for small- to mid-sized fabless companies. These companies must overcome three major hurdles to effectively close the loop with their manufacturing chain: data sources, infrastructure and collaboration tools.

11.3.1.1 Data Sources

Successfully managing manufacturing information is a challenge for the fabless business model. First, gathering the widely disaggregated sources of manufacturing data, a consequence of the outsourced manufacturing business model, is difficult, particularly in an environment where standards for data exchange are largely absent. It is the norm for a fabless manufacturing chain to have multiple foundries and several test suppliers for wafer sort and final test with file formats ranging from comma-separated text files or XML to binary STDF. Data files may range in size from 50KB to 50MB or more; too large to be loaded into standard commercial spreadsheet products for analysis. An additional complication is that some fabless companies test parts internally, generating a combination of STDF files, bench data or image files created during failure analysis.

Data quantity, complexity and quality further complicate the picture. As feature sizes continue to shrink, foundries will be required to more tightly control parametric and in-line error budgets, such as critical dimension. The result will be that design- and parametric-related yield issues will become an increasingly larger portion of the overall yield pie. This trend will greatly increase the quantity and complexity of data required to identify and resolve yield and performance issues. Data quality issues involve inconsistent lot naming, file corruption, lot splits and merges, engineering lots, bench data, data verification and cleansing (Figure 11.5).

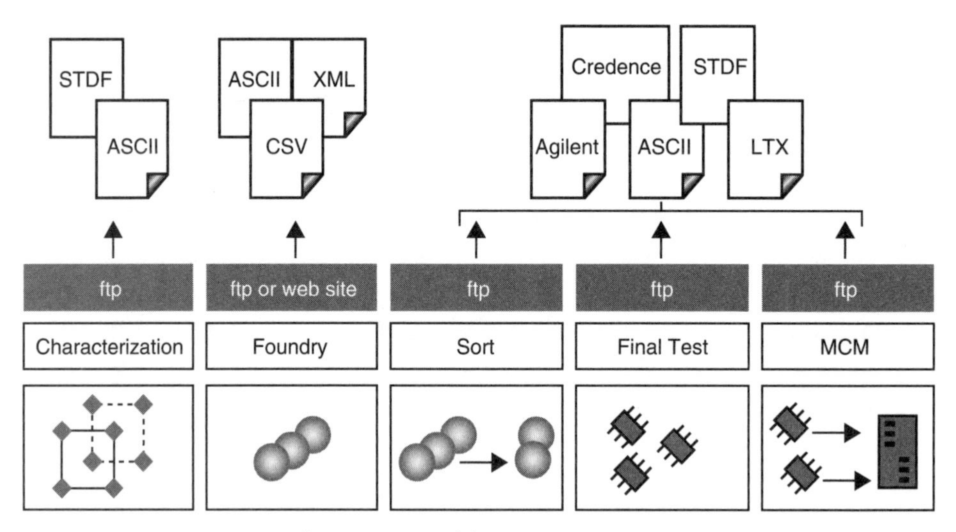

Figure 11.5 Fabless Data Sources

11.3.1.2 Lack of IT and Analytical Infrastructure

Most small- and mid-sized fabless companies focus their infrastructure investments on directly supporting their design activities; indeed, that is the reason these companies exist. In addition to analysis of yield and production data, product engineering and operations staff must also download, cleanse, transform and merge data from suppliers.

Commercial spreadsheet products are often used for data analysis. However, these programs have restrictions on the number of available rows and columns, which limits the amount of data that can be analyzed. An additional problem with commercial spreadsheets is that they do not support semiconductor analytical requirements such as wafermaps, boxplots or other sophisticated analyses.

Some fabless companies have or may be considering deployment of a semiconductor yield management software (YMS) product to address their data analysis needs. Most commercial YMS products on the market today were designed to meet the data and analytical needs of the fab. They poorly provide, if at all, the conditional analysis or part and pin-level support required by product engineering. In addition, fabless companies must consider the additional infrastructure costs tied to hardware, security and system administration; as well as the time required to install, configure and integrate these systems – some of which may take 3 to 4 months or more to become operational.

11.3.1.3 Collaboration and Organizational Learning

An often undervalued strength of the IDM business model has been the close interaction of design and product engineering with manufacturing and process development. Where it exists, this intimate collaboration helps close the loop and promotes improved understanding of design-process interactions.

A fundamental challenge to managing the "arms-length" interface between a fabless design firm and foundry is coordinating the smooth and efficient transfer of new designs into full-volume production. Fabless companies, by their nature, must collaborate across company boundaries to coordinate development tasks and exchange information to support product and process development. While strict adherence to foundry-specified design rules and process flows with access to selected PCM and in-line data can mitigate several issues, solving yield and product problems requires timely exchange of data, analysis and information from all elements of the manufacturing chain, such as foundry, wafer sort, assembly and final test.

11.3.2 Outsourced Device Characterization and Production Monitoring

Dealing with manufacturing data, systems and analytical infrastructure and collaboration issues can channel a significant portion of an engineer's time into less-productive activities. These challenges, exacerbated by time-to-market pressures from management, partners and the investment community, are leading many fabless organizations to outsource the hosting and management of their production monitoring and device characterization environment. Figure 11.6 summarizes the different drivers for outsourcing a company's component verification and monitoring solution.

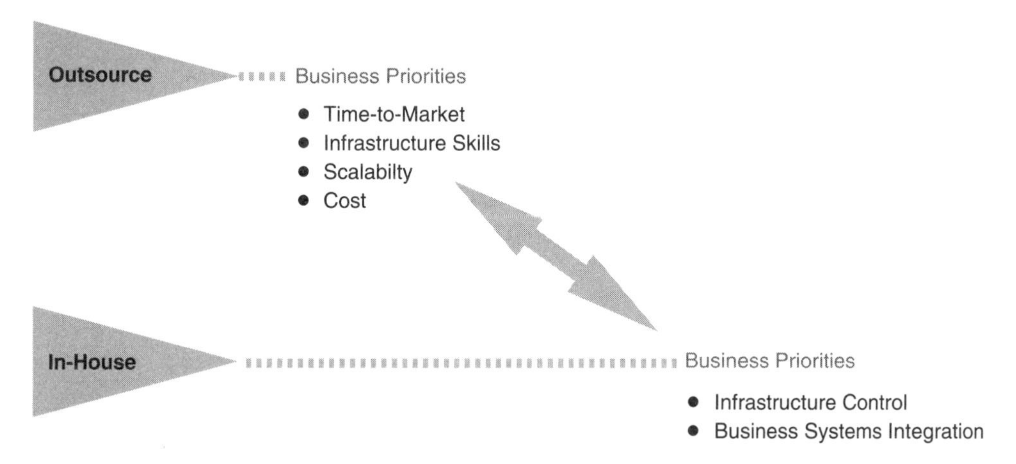

Figure 11.6 Internal versus Outsourced Hosting

The widespread adoption of the Internet, modern browser technology and the stability of commercial hosting models with guaranteed availability makes it possible to launch a company-wide production monitoring and reporting system in a matter of days with little or no impact on a company's IT infrastructure.

The business priorities of small- to mid-sized fabless companies tend to be dominated by time-to-market, cost control and lack of infrastructure. In general, these priorities lead a business to outsourcing to meet their requirements for manufacturing data management and analysis. The ideal solution would automatically sweep data files from the .ftp sites provided by their manufacturing suppliers then validate, cleanse, translate and insert both raw and summarized data into a secure data mart. The data mart would provide secure data access and be highly scaleable to meet growth requirements in terms of data and numbers of users. One company uses a similar data mart set-up and has stored more than 2 years of wafer sort data only in excess of 300GB with more than 100 engineers accessing the system for reports and analysis.

The data mart would allow engineers to use standard desktop tools. For example, standard Internet browsers such as Microsoft® Internet Explorer would allow users to view and filter their manufacturing data on the fly; create automated reports and alerts; and correlate operation, lot, wafer, die, site, part or pin-level data from anywhere using secure access on the Internet. A widely available, browser-based environment would then help to increase the level of collaboration by allowing report and analysis results to be captured and shared within the company or outside with suppliers and customers.

11.3.3 Summary

In today's fabless environment, the best design is not always the most successful one. At the end of the day, the race will go to those fabless companies that develop a clear understanding of their manufacturing chain and its effect on their product. Outsourced solutions for the rapid detection of yield and production issues provide fabless semiconductor companies with an avenue to monitor the status of their products through their entire manufacturing chain; enhance their level of collaboration with their manufacturing, test and assembly partners; and close the loop by cycling vital production data to their designers. To enable this model, engineering analysis tools must be simple, easy to use, relevant to product engineering and have minimal impact on infrastructure (Figure 11.7).

11.4 Managing a Virtual Manufacturing Chain

No fab? No problem. Manufacturing execution systems (MES) fits the foundry model as well as the traditional fab, with the added benefit of managing a virtual manufacturing chain. The need for MES software in a successful semiconductor fab has been demonstrated since the mid-1980s. MES is integrated with equipment and automation to control the complex manufacturing process, and to provide complete traceability of multiple products moving through multiple process routings.

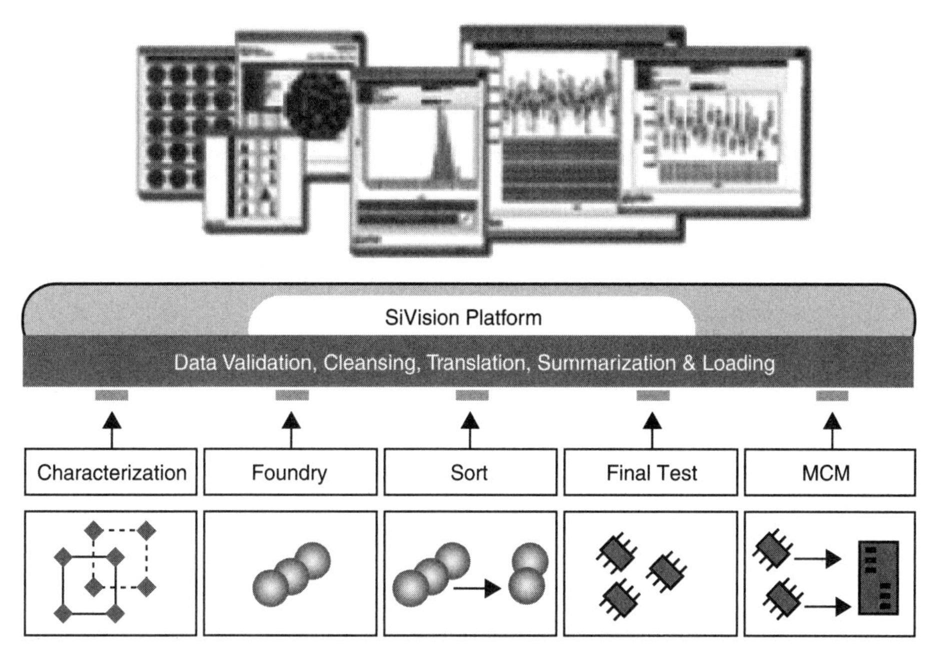

Figure 11.7 Platform Example

In the end, customers do not care if the product is manufactured in a captive fab or a foundry. But they do care about on-time delivery, cost and the reliability that comes from complete process traceability and certification. Fabless companies that rely on the foundry model must manage and control production at an increasing number of global subcontractors, and provide complete traceability of the virtual manufacturing chain. Consequently, fabless companies shipping first product and those with up to $1 billion in revenue are realizing competitive advantage by utilizing MES.

Business applications that offer accounting and billing functions do not necessarily have the flexible manufacturing capabilities needed to manage a virtual manufacturing chain. However, MES systems that are designed to manage fab, test and assembly production do have those necessary features. Consequently, MES is being used as part of a solution that also includes supply chain Advanced Planning and Scheduling (APS), as well as business applications such as ERP.

11.4.1 Information Equals Success in a Fabless Environment

Success in the fabless market depends on offering multiple well-designed products that meet the needs of a changing market, and on managing multiple subcontractors that form a virtual manufacturing chain. Effective virtual manufacturing chain management includes:

- Securing and utilizing foundry capacity
- Consistent on-time delivery
- Tracking product and subcontractor performance and cost
- Maintaining complete process traceability

Information is the key to taking control of these factors, and MES provides effective communication with all manufacturing chain participants including customers. The challenge is to keep track of all subcontractor production as material moves through each segment of the chain. To address this, MES tracks lot status information of all subcontractor production activity on a daily, or more frequent, basis. MES also provides real-time reporting of lot status throughout the virtual manufacturing chain to help fabless planners and their customers better manage production.

Effective use of MES for fabless production is possible because of new technologies such as electronic data interchange (EDI), the Internet and emerging standards such as SMDX. These technologies offer means to automate subcontractor data collection with a frequency and volume that is required – and that is not feasible using previous fax, e-mail and manual data entry methods.

11.4.2 MES and APS Drive On-Time Delivery

On-time delivery is one of the most critical success factors in fabless manufacturing, yet it is one of the most difficult to achieve. Customers who practice JIT (just-in-time) manufacturing request delivery one day before a production build. How do fabless planners ensure a one-day-early/zero-days late delivery on a commitment 3 months in the future?

An MES system integrated with an APS package allows a virtual manufacturer to optimize both halves of the "commit-hit" equation. Fabless planners use APS with real-time MES data to quote delivery quantity and date. Planners then use MES to manage virtual production to deliver on that date, and they do it with a consistency that helps to achieve "preferred vendor" status. Realistic commitments are possible because APS systems utilize MES data that indicates the manufacturing chain's ability to deliver. This data includes:

- Present yield and cycle time variance
- Historical yield and cycle times
- Lead times for various subcontractors

- Committed capacity for various subcontractors
- Current status of all lots in the manufacturing chain

When a customer calls asking for a delivery commitment, the request is entered in the APS model and "what if" scenarios are run based on various quantities. The output is a delivery date that has an acceptable probability of being achieved.

Meeting those commitments means repeatedly checking production yield and cycle time to determine which orders are on schedule, and taking timely action when orders are behind schedule. Immediate access to the status of all orders in the manufacturing chain is critical for managing on time delivery.

Without combined MES and APS capabilities, fabless planners either sandbag inventory or under-commit quantities to guarantee on-time delivery. Both are undesirable alternatives, since excess inventory is expensive and competitors may be filling the remainder of an under-committed order. Without MES, planners are unaware that orders have yield or cycle time variance, until a delivery commitment is missed or until it is too late to solve the problem. It takes only a few missed deliveries to lose preferred vendor status with the best customers.

With timely knowledge of a subcontractor's yield and cycle time variance, it often is possible to start an additional lot or to expedite production to make up for the problem. Even if a delay cannot be overcome, the customer can be notified immediately to re-schedule their JIT production. The combination of MES and APS allows fabless planners to optimize both halves of the "commit-hit" equation, and also provides fabless companies a strategic advantage over those with less capable tools.

11.4.3 MES and ERP Drive Subcontractor Management

An ERP package with a general purpose manufacturing module may not provide all of the required fabless manufacturing functionality. Although ERP does an excellent job of handling financials and order management, it does not provide the real-time data collection and flexible lot tracking features needed to manage a virtual manufacturing chain.

An MES system integrated with an ERP package allows a virtual manufacturer to effectively manage global subcontractor production. The ERP package is used to open subcontractor service and material purchase orders. MES is used to track wafer lots through fab, individual package lots and sub-lots through assembly and test, and to tie real-time subcontractor activity back to ERP purchase order and financial functions.

With virtual manufacturing, the exception often is the rule. Production planners regularly deviate from plan in order to address changing subcontractor performance and changing customer demands. The flexible MES tracking functions needed to manage this dynamic environment include:

- *Real-time lot status*: provides a daily snapshot of all subcontractor WIP activity as well as yield and cycle time variance.
- *Splits and combines*: one lot can be split and have a portion combined with another lot to replace unplanned yield loss.
- *Downgrade*: higher performance parts can be split from a lot and downgraded to a lower performance part number to fill an order.
- *Binning*: one lot can be binned into multiple lots and part numbers after package testing.

In addition, an MES system with an activity based costing capability allows more effective subcontractor management. MES can allocate actual costs to individual products and subcontractors, perform monthly inventory valuation and roll that information into a general ledger ERP function. That means using MES to track production activity, and to complete yield and cost traceability as a wafer lot converts to package lots and sub lots; and through splits, combines, downgrades and bins.

Without MES integrated with ERP financial capabilities, planners cannot tie real-time subcontractor performance back to service and material purchase orders. That linkage is critical and is not available with a stand-alone ERP system.

With the combination of purchase order and lot-level performance data, planners can allocate costs to actual subcontractor performance. This information is used to determine the profitability of subcontractors by product and helps with future capacity allocation decisions. The combination of MES and ERP allows fabless planners to better manage subcontractor performance, and provides fabless companies a means to profitably manage the virtual manufacturing chain.

CHAPTER 12

Building the Right Partnerships

12.1 Suppliers are (Almost) Just as Important as Customers

At a supplier's symposium event, Dr. Santanu Das, CEO of TranSwitch Corporation, said that "suppliers are almost just as important as customers." Dr. Das' remark clearly conveyed a key important message to his fabless semiconductor partners: "we are all in this together and we can only succeed by working closely together."

Clearly, without the right partnerships, a fabless semiconductor company cannot deliver its products to its customers, and as a result, such customers would eventually be lost, thus the need to build the right supplier–customer partnerships. This chapter provides a checklist of aspects to consider when selecting a supplier, both from operational and legal points of view. This chapter also has a case study on semiconductor electronic design automation subcontracting which highlights the issues to contend in the other side of the partnership, that is, from the point of view of the supplier.

12.2 Operations in a Fabless Start-Up

12.2.1 When to Evaluate Supplier and How to Go About It

Define Place in Design Cycle

- *Product definition*: Look at all fab suppliers who have processes that meet performance and IP requirements.
- *Register-transfer-level (RTL)*: Narrow down the list to those with capacity, and those with an interest in your business. Make sure you can meet financial performance based on cost modeling.
- *Netlist*: Begin evaluating assembly suppliers with capacity and capability to do the package which meets technical and cost requirements. Make sure they have engineering support to meet your requirements for substrate and qualification.
- *Tapeout*: By tapeout, all suppliers need to be finalized, including test and making sure the appropriate supplier has capacity and the capability to meet wafer sort and final test needs. This should include load board and probe card design and development.

Establish Process and Criteria for Supplier Evaluation

- Based on the company structure, executives should decide who should be part of the supplier evaluation team and who the final decision makers are.

12.2.2 Build Your Supplier Requirements Matrix

Standard Process Requirements

- Design engineering needs to define the type of processes most of your products will need; for example 0.13-micron CMOS in a low-voltage process.
- Product engineering needs to compile a list of probable testers and configurations needed to compare against.
- Assembly, engineering and marketing will need to compile lists of package body types/thicknesses/pin/ball spacing, material requirements and design capability.

Custom Process Requirements

Design needs to develop a list of process performance requirements. If this is not met by a standard foundry process, alternatives need to be evaluated and priced out if they add to the base wafer price.

IP Requirements

- Design engineering should establish acceptable hard IP requirements, and if they need to be silicon-verified at the supplier in the process being used.
- Compile a list of IP needed from the foundry supplier directly, review the IP in the supplier process for number of mask layers needed as a potential for cost impact.

Information System Requirements

- Define work-in-process (WIP) and update requirements (daily, real time, format to directly upload into your WIP system).
- Define product wafer sort and final test yield format and frequency of transmission.
- Process, assembly and design engineering need to define online data required on the suppliers' websites.

12.2.3 Supplier Evaluation Team

Engineering Role

Evaluate suppliers' IP qualification data, review yields, process qualification data.

Operations Role

Review IT capabilities and communication standards, evaluate the customer service team, meet all executives and present your company's story.

Quality's Role

Review the supplier's process control data, internal quality processes for corrective and preventive actions, meet and evaluate the quality team.

12.2.4 Find the Right Hook to Sell Your Supplier for Choice

Determine What You Have to Offer the Supplier

- In the first executive meetings present company management, product, target market, committed customers, competition, funding, forecast and capabilities.
- Listen carefully to their questions. If there are no questions, you are missing the right content.

Determine What the Supplier Needs or Wants

Determining what the supplier needs or wants for the business is important as it will provide a good idea of the potential fit for working together. For example, it may be that the supplier is trying to diversify from its heavy dependence on one specific market, say consumer electronics or automotive, into new emerging markets. If this is the case and your business is new and emerging, then you will find a good reception even though your company is new and small. The converse is also true. If the supplier has already diversified into your type of market(s) and/or its business plan is totally focused on one particular market segment or customer base, then you may have a harder time convincing the supplier to work with you.

Put Together a Plan to Sell the Supplier on Your Company

- Define who will talk to each level within the supplier company and the frequency with which you need to get their buy-in and support.
- In the product development stage, don't do a hard sell. The time to pull all the stops is when you have working prototypes and a real customer.
- Have operations provide quarterly updates while in early development.
- Close to tapeout, start providing a monthly forecast and update.
- Once prototypes are working, provide a monthly update of customer testing results, characterization and qualification.

Sell, Sell, Sell

- The CEO should definitely meet and present to the key executive of the supplier in their headquarters after working prototypes are complete and ready to go to production.
- The more people within your suppliers' organizations that you can sell your company, people and product on – the better.

12.2.5 Establishing Communication Channels and What to Communicate and How Often

Defining Who Should Communicate with Suppliers

- Create a list with all possible areas for communication with suppliers and define the primary and secondary contacts within the company. At a minimum, areas should include: purchase orders, build plans, lots on hold, discrepant material, shipments, invoicing, technical requirements and business negotiations. Establish who the decision makers are and who has final say should there be a disagreement.
- Include information for name, work e-mail, work number and an after-hours number.
- Have the supplier complete the corresponding information.
- Update the list at least every quarter, and review it as part of your quarterly business review.
- Establish how purchase orders will be placed; blanket by product or package type or lot by lot to go from purchasing or planner to customer service at supplier.

Daily

- Obtain daily updates from your fab, assembly and test suppliers for any lots on hold that require disposition. This information should be directed to the appropriate test or product engineer and planner.
- Expect daily ship alerts from your suppliers to your product planner and any pass any changes to ship or build requirements from your planner to your suppliers.

Weekly

- The planner or buyer should deliver weekly build, test and ship plans to assemblers and test houses (if different). You should ensure this coincides with the assembler's weekly factory scheduling.

- The planner should send assembly build requests to the supplier factory for specific lot builds.
- The planner should deliver the die start schedule to the fab.

Monthly

Planning, purchasing or operations management should deliver a 3-month forward-looking assembly, fab start or out and test plan. Ensure you will not exceed credit limit or start working that early.

Annual

You should do an annual plan for delivering the following year long-term forecast for fab suppliers per their annual planning cycle. This should be broken down by fab process/ technology and any sort requirements.

Quarterly Business Reviews (or Semi-Annually)

- Have the company agree to a supplier score card. This should include input from manufacturing, process, assembly, product and quality engineering, planning, procurement and design engineering.
- Senior management should attend if you expect senior management from the supplier's side. They typically will match job titles. Kick off with a business update, including new customers, products and company performance.
- Quality should present any customer or product discrepancies and returns.
- Engineering should present product yield performance and any product/test/package changes planned.
- Planning should present the upcoming quarterly build plan and delivery performance.

12.2.6 Know When and How to Negotiate

- Most pricing for wafers is by product or by technology.
- Once a product yield is stable (approx 500 wafers), evaluate negotiating for die buy pricing. This occurs when the foundry supplier takes ownership for the yield.
- When you are selling the supplier on your business and you have no volume, this is not the time to do hard negotiation if you are in a capacity shortage situation. Use modeling to determine an acceptable range for die, assembly and product test costs.

- Working prototypes, with firm customer orders is a good time to start real negotiations. Negotiate for time-/volume-based pricing.
- Spot buy pricing for a sudden up-tick in a specific product demand should be taken advantage of. This only works in an excess capacity market.
- Determine who makes all pricing decisions and know the negotiations will go several rounds. You need to have a matching level of executive negotiating with that person at the supplier base.

12.3 Legal Issues for Fabless Semiconductor Companies

Below are some legal issues that a fabless company should consider starting from its inception.

12.3.1 Incorporation Basics

A. Which form of association should the company be established as: a "C" corporation? Subchapter S tax treatment of a C corporation? LLC?

If either a C Corporation or an LLC are selected, you will need an agreement among your founders that determines who does what and who gets what and how a founder can be removed and what happens if a founder shareholder dies or is permanently incapacitated, etc. In a C Corporation this is called a Shareholders' Agreement and in an LLC it is an Operating Agreement.

B. Where do you form your entity? Your home state? Delaware? Nevada? This depends on your purpose in forming your company and on your exit strategy.

12.3.2 Protection of Your Intellectual Property

A. Rely on dedicated and relentless use of non-disclosure agreements (NDAs) with all IP.

B. Make sure you do patent filings from the beginning.

C. As a preventive measure: review the US Patent Attorney Office Internet Site and search for all registered patents in your area. Look for patents that could block your new technology and engineer around them so that you do not infringe all elements of the dangerous claims. Don't wait for the patent owner to discover your product. BE PROACTIVE.

D. Determine whether you will use subcontractors or consultants to help you design your product. If you do this, you MUST enter Consultancy Agreements with an appended Assignment of Inventions that forces the consultant to make you the sole owner of the consultant's work product created to your specs.

E. Establish an Intellectual Property Committee at your company that reviews with patent counsel and senior engineering management what your company is developing and whether there are patentable innovations and that reviews third-party IP challenges. Meet regularly and keep a record of your deliberations and action assignments.

12.3.3 Foundry Relations

A. Establish foundry relationships that can be trusted.

   1. Enter foundry contracts that protect your parts by way of NDA and Non-use and an IP indemnity that covers the fabrication of your part (but not your design.) Get an agreement as to lead times and how the fabrication center deals with assigning priority of fabrication to its customers should the fab center be overloaded. Don't wait until they are overloaded and you are without supply.

   2. NRE assigned to your foundry: make sure it is clear in the contract that you own the IP created by paying for NRE, no different than if you were contracting with a design consultancy house. Do not let your purchase order try to do all this work.

B. You may want to have discussions with other foundries so that if you decide to change foundries, you are familiar with those who can manufacture your product with the least amount of disruption your design flow.

12.3.4 Legal Services

1. Establish a close relationship with a lawyer that understands your technology and business model. Let legal counsel guide your overall legal services and meet on a regular basis with management.
2. Establish a close relationship with a patent lawyer who has a good background in your technology and who can and will understand your innovations. Establish a strategic connection between this patent lawyer and your Intellectual Property Committee. Let the Committee dominate what gets patented and how broad the patents should be from a strategic planning viewpoint.

12.3.5 Human Resources

1. Establish a Human Resources (HR) function as soon as possible.
2. Establish HR policies regarding corporate ethics, observing the law, etc.
3. Train your employees in honoring third-party IP restrictions, not violation US law that protects others' IP, etc.

4. Don't hire illegal aliens, no matter how inexpensive.
5. Make all employees sign an Assignment of Inventions with a 1-year covenant against competition that is written to be enforceable where the employee works, a covenant against soliciting your employees to leave your company and against soliciting your customers for 2 years.

12.3.6 Standard Contracts

Early on, establish standard contracts for each category that your early start-up will encounter:

- NDAs
- Technical Consultancy Agreements
- Assignment of Inventions for technical consultants
- Assignment of Inventions with Restrictive Covenants for employees
- Patent Disclosure Form

12.4 Semiconductor Back-End Subcontracting: No Longer a Zero-Sum Game

The scenario in which a backlog of orders exists in excess of manufacturing capability is an age-old manufacturing quandary: to subcontract or not to subcontract. To answer this, the manufacturer must determine if the backlog justifies investment in additional resources, and determine if time permits to attain those resources. If this case exists, then the business is generally subcontracted. The manufacturer, by using the subcontractor, gains the required capacity, and the subcontractor gains the business – a win-win situation. However, when the additional capacity is no longer necessary, the business is often pulled back inside, resulting in the subcontractor losing the newly gained business. In this scenario the customer also becomes a competitor, which results in a zero-sum game.

Today, however, that model has changed. The subcontractor is now an integral part of the semiconductor supply chain. The new model has an end in mind that is being adopted by the entire semiconductor industry. By most surveys and analyst estimates, outsourcing will grow faster than the semiconductor market in general, indicating how crucial subcontractors are to the success of most semiconductor companies. Manufacturers without back-end operations, that is the fabless and chipless system houses, really have nowhere else to go. Even integrated device manufacturers (IDMs) are aggressively cutting spending on non-value-added assembly and test activities in favor of outsourcing. Since the subcontractors have developed advanced packaging technology beyond what is internally available, most manufacturers have curtailed investment in new packaging technology, further increasing their dependence on the subcontractors.

Subcontractors are paid per device and invest incrementally in capacity, resulting in high equipment utilization and lower fixed costs – yet they may actually adopt a less efficient but more flexible manufacturing process than IDMs. This model meets manufacturers' changing needs and peak requirements. By running high volumes of certain packages used by multiple manufacturers, the subcontractor's overall costs are lower, and the reliability and quality are often higher. Strong production processes translate into high cumulative yield. Although very attractive to the manufacturer in terms of reduced costs and risk, the new logistics can become complex and difficult to manage.

To mitigate this, subcontractors have developed advanced data and information management services online via the World Wide Web. They manage the logistics of die in and packaged devices out. Their operations are designed to mirror the manufacturer's internal systems and processes, thus providing a virtual factory environment. It is now clear that the manufacturer's association with the subcontractor has gone beyond a classic vendor/customer relationship and has even surpassed a synergistic one. It has, in fact, become symbiotic and is no longer a zero-sum game.

12.4.1 Turn-Key Processes

More and more, the subcontractor is asked to provide turn-key service that allows manufacturers the opportunity to consolidate all their back-end related needs from package design to wafer probe, assembly and final test. An additional service is the managing of finished goods inventory, packing, labeling, storage and shipment to the manufacturer or end customers. Another strategy employed is the "pull-and-pay" process, where the manufacturer supplies wafers or die to the subcontractors who package them at their convenience, thereby fully utilizing their equipment. The finished devices are warehoused, and the manufacturer only pays when requests are made to pull them from inventory to ship. As shown in Figure 12.1, multiple entry points to this flow accommodate manufacturers who prefer less-than-full turn-key services.

Because the majority of the devices assembled are tested internally by the manufacturer, testing has become both a strategic opportunity and an asset for most subcontractors. Subcontract testing has the same economic benefit as package assembly. Often, for security reasons, or because test is considered a strategic advantage, it is kept in-house. Test can reveal the internal workings of the chip which is generally proprietary. In other cases, the manufacturer feels the subcontractor has limited test knowledge and capability to resolve problems since they do not have the clear visibility into the functionality of the design, creating a Catch-22 situation. Still, subcontractors are providing more integrated services,

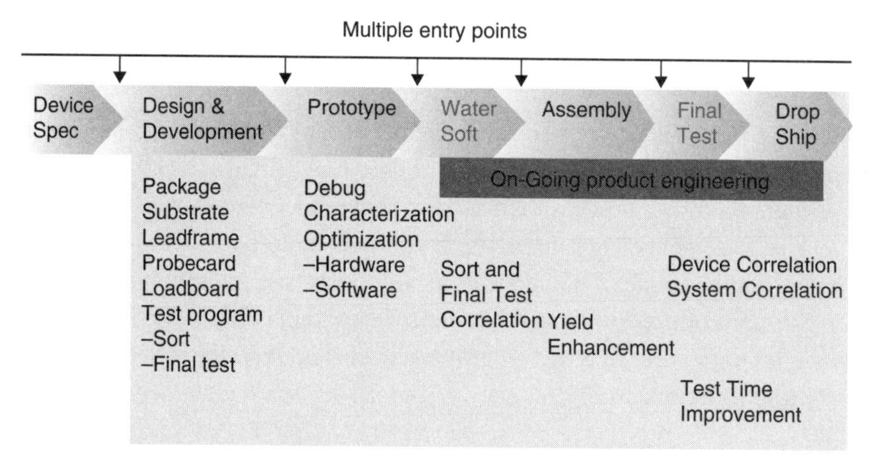

Figure 12.1 The Turn-Key Model

including test, as the issues around security and test capability quickly dissipate. If the manufacturer is to remain competitive, the cultural readiness to set aside the old for the new is mandatory. The added value of testing in the location where the product is assembled is obvious.

12.4.2 Customer Engineering

To meet the fast design cycle times and quick device qualifications demanded by the manufacturer, the subcontractor needs both strong process and product development functions. Often some customization with respect to technical specifications and materials is required. Figure 12.2 is a typical customized package configuration. The package design engineers

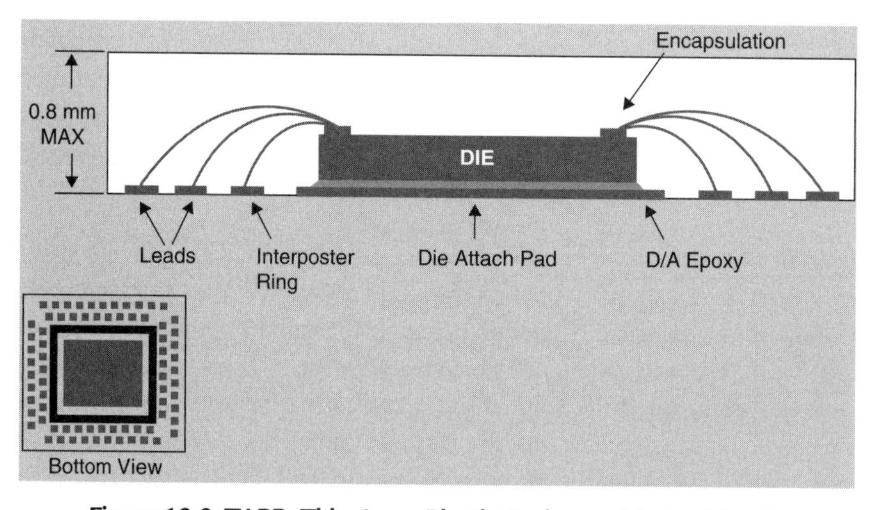

Figure 12.2 TAPP: Thin Array Plastic Package with Gold Leads

work with the manufacturer to develop application-specific solutions using their standard product base. Having both factory and local design interconnection, computer simulation and modeling centers speed time-to-market. Employing thermal and electrical modeling services before building a new design saves time and money. Providing the additional service of thermal and electrical verification testing on design models significantly improves the chances of achieving successful results.

Since the package has become an integral part of the functionality and performance of the chips, final production yield is often a function of integrating the package design with the test strategy. Test programs are developed at chip design time, and this requires the integration of the packaging and test services. Thus, where historically the twain shall never meet, this new paradigm makes the introduction and ensures the resulting connection between packaging and test services.

12.4.3 Customer Service

The customer service function manages the production, the flow of wafers or die in, then the shipment of packaged devices out. Weekly, if not daily, demand forecasts from customers are translated into factory output requirements. The forecast is used to both buy materials and plan capacity. Without die, the subcontractor expires. The lost capacity can never be made up; therefore, it is critical to employ accurate forecasting models that reduce costs and speeds product delivery.

The customer service personnel are also purveyors of information and analysis including inventory position, cycle time history, on-time delivery performance, shipping information, statistical process control and yield reports. Quarterly business reviews are conducted to analyze these reports and to allow the manufacturer the opportunity to rate the subcontractor's performance versus commitments and expectations. This results in corrective action or improvement programs. The bar is never lowered as they asymptotically drive for perfection.

12.4.4 Customer Relationships

Open lines of communication and an active exchange of ideas are keys to the success of this relationship. Proactively confronting and resolving the differences of opinions and misunderstandings that invariably arise in any close association strengthens the rapport. If the objective of the subcontractor is to be a high-volume, low-cost, mammoth supplier of the world, those interests may not be aligned with the majority of the manufacturers. The fabless and chipless varieties especially need personal service in addition to low cost. It is here that overall cost of ownership and the value added by the subcontractor come into play. Partners must be financially and operationally stable with a common vision and strategy. They should be philosophically aligned, and selectivity is important to both parties. They must be committed to continuity and the success of the relationship. The question of how much to give and how much to expect in return is one of the ongoing challenges.

The subcontractor will educate the manufacturer about various solutions to their back-end requirements. The products and services offered are viewed as the facilitator of the solution and not the solution itself. As skepticism withers in the face of success, this is the time they develop a mutual commitment to the solution. The subcontractor, with an understanding of the manufacturer's end customer, actually suggests strategic packaging direction and opportunities. The value of this relationship is increased market share as well as a competitive advantage for both. The subcontractor, in effect, becomes more than a vendor to the manufacturer – the subcontractor has become a virtual extension of the manufacturer's business in an enduring symbiotic relationship.

CHAPTER 13

Building the Right Relationships with the Board and VCs

This chapter discusses the importance of building the right board of directors and maintaining good relationships as the company grows. It will also discuss the value that venture capitalists (VCs) bring to the fabless company from a network and relationships points of view in addition to the venture capital that they bring.

13.1 Creating Successful Corporate Boards in Fabless Companies

13.1.1 How Effective are Corporate Boards in the Semiconductor Industry?

Although many semiconductor companies, both fabless and integrated device manufacturers (IDMs), currently possess highly effective boards, many semiconductor company boards probably would not make anyone's "top ten" list.

Heidrick & Struggles (H&S) and FSA performed a joint study and completed a series of interviews with several semiconductor industry board chairmen. The study was aimed at building a body of knowledge about how semiconductor industry boards operate and the differences, if any, between boards in fabless companies and IDMs.

The study produced high-quality information, much of which was not surprising. However, the study produced several unexpected findings. One of which was that many semiconductor company board chairmen wish their boards could have greater impact on the success of their companies.

13.1.2 What do Effective Boards Look Like?

Several years ago, a board consultant named Ram Charan wrote a book titled "Boards at Work – How Corporate Boards Create Competitive Advantage." In a chapter titled "What Good Boards Do," Mr. Charan described an effective board as one that is a "vibrant participant in the corporation's value creation process."

Viewed against that standard, many semiconductor company boards must make several changes if they are to become real participants in creating value. What should they do, and how can they do it?

Companies must attract and retain board members who have sufficient time available to do a first-rate job.

Having ample time would seem like an obvious "must have" characteristic for successful board members – but often, board chairmen choose members who have too many other commitments. Take a look at semiconductor company board rosters. They're peppered with sitting chief executive officers (CEOs) of other companies, or "retired" executives who sit on five, six or more boards. Typically, start-up companies also compete for attention from VC firm partners who sit on the boards of multiple other start-ups.

The FSA semiconductor industry study found that board chairmen believe their board members need to devote between 2 and 6 weeks per year to do an effective job. This is far more time than board candidates typically expect will be required – and it is a lot more than many current board members actually spend. In the post-Enron environment, the time required of board members will probably increase beyond the 2–6 week estimates the board study participants made several months ago.

Companies should expect and demand that their board members be deeply involved with the company and with the board.

In the past, it's been easy for some board members to get by on superficial involvement – they show up for meetings four times a year, listen to a series of carefully staged presentations, give their affirmative vote to management's proposals and have a pleasant dinner with other board members before heading home. The time when that was acceptable has passed. The best board chairmen are now changing the rules. They expect their board members to get actively involved with the company's business. They are holding more extensive and inclusive board meetings, with formats designed to force members to get beyond the superficial levels of discussion that have been the norm. They are requiring executive and line managers to interact in depth with the board, and to educate the board in the nuances of the organization and the changing competitive environment. They are constructing active programs to acquire and keep board members involved with the company, acquainted personally with the officers of the company and in touch with key customers and stakeholders. They're making board membership more of a job and less of a position.

Boards should be balanced so they include the best set of experiences and skills, given regulatory requirements and the limits of the talent pool.

Boards need members who have expertise in marketing, finance and technology. Knowledge of the company's particular business niche is important, but so is having familiarity with the business of its customers, and even its customers' customers. Just recently, board chairmen have an entirely new set of requirements to consider regarding financial knowledge, unblemished records, absence of conflicts-of-interest and separation from management duties in the company. In some situations, specific regional or international backgrounds are needed on the board. In other situations, boards need members who can open doors to customers, capital sources or regulatory bodies.

There is often a need for boards to include representation from specific organizations or population segments. With all these factors to consider, board chairmen still need to assemble collegial groups of individuals who can work well together, make decisions logically and quickly and provide sound advice and mentoring to both management and the chairman. Now, many previously eager board candidates are far more hesitant to accept board seats. Some of them are wary of the increased liability; others want to avoid conflicts-of-interest; and many simply don't feel they have the time that's needed in today's business environment. The net of this is board chairmen need to devote much more effort and greater care in finding and recruiting board members.

Companies should work toward separation of board leadership – having the chairman and the CEO roles held by different individuals.

Ralph Ward, in his monthly publication, "Boardroom Insider," provides timely observations and insights on developments affecting corporate boards.

In a recent issue, Ralph noted that the trend toward splitting the chairman and CEO roles is gaining momentum. He said, "The combined CEO/chair role has long been one of those American business oddities that puzzle the rest of the world. The conflict of a CEO leading the body that grades his report card is obvious. The growing momentum toward splitting the chairman and CEO roles between two individuals is a positive development, and over time should make a real difference in board effectiveness."

The days when the role of chairman and CEO could be combined in one individual are numbered. Separating the roles will create difficult challenges, but those challenges can be overcome. In the long run, this will be the pattern for successful semiconductor company boards.

13.1.3 What About Fabless Companies?

All of the above rules apply to fabless companies, as well as any company, whether public or private, large or small, domestic or global. But are fabless companies special cases? Are their business models so different from other companies that they need their own set of guidelines for effective boards?

The answer is yes. Fabless companies are different. A biological metaphor may help to illustrate the difference. Larger companies, including most IDMs, are like mammals, while fabless companies are like birds. Mammals can go for relatively long periods without eating, because their robust skeletal structures enable them to store nourishment. Birds, by contrast, must stay light to be able to fly. They need to eat often, and they have few reserves to call on if food is not available.

Fabless companies, especially start-up companies, are in the bird category – many are more like hummingbirds. They must keep moving every minute, and their margins for error are exceedingly small.

Fabless companies need board members who understand this characteristic. Fabless company boards must assimilate information quickly and react rapidly. They must be attuned to the market; have close relationships with their foundry and assembly/test partners; and have multiple links to sources of capital. All semiconductor company boards should be quick on the uptake and spring-loaded for fast action. Fabless company boards just need to be a bit faster with somewhat tighter springs. Beyond these, successful fabless semiconductor company board members should include the following:

- Deep knowledge of the particular product and/or market area the company is in, or wants to be in
- A strong passion for overcoming difficulties and coming out a winner
- A willingness to get deeply involved and to do whatever is needed to ensure the company's success
- The ability to spend major amounts of time and effort coaching, problem solving and helping to make things happen
- Jackrabbit reflexes, and an ability to make quick decisions based on little hard data
- A disdain for formality, pomposity and traditional board trappings
- A built-in sensitivity to the company's changing needs, and a willingness on the part of each board member to leave the board voluntarily if their skills or ability to contribute no longer fit the situation.

13.1.4 Summary

The most successful executives possess large quantities of passion, commitment, appropriate skills and personal drive to succeed. Successful board members need the same characteristics. Board success is a function of the people who make up the board and the person who runs it. Find people with appropriate skills, sufficient available time, a passion for success and the right mix of collegiality and independence. Put them with a chairman who leads, involves everyone and insists on active participation and measurable contribution, and the outcome will be a great board.

13.2 Finding the Right VC

At the end of the 20th Century, entrepreneurism had reached epidemic proportions, and venture capital (VC) firms, both established and new, competed for the best investment opportunities. In the first few years of the 21st Century, the epidemic has abated and VC firms have become much more selective in their investments. In the current economic environment, it is tempting to assume that any money is good money. That can be a risky assumption.

Public perceptions about the investment climate tend to be overstated. It is a misconception, for example, that "anyone with a good idea" could be funded at the peak of VC investments in late 1999 and 2000, according to Guy Hoffman, Venture Partner with TL Ventures in Dallas. Conversely, it is also a misconception that no one is being funded today. "VC financings are occurring each and every day. The VCs have merely tightened up the definition of what constitutes a good deal. If you take out the 2000 "bubble," VC investments are still on a positive curve," Hoffman said.

Thus, while entrepreneurs may no longer be able to play one eager VC against another to negotiate favorable terms, many VC firms have substantial "dry powder" and are willing to invest it in the right circumstances. Even in today's economy, companies with good business plans and talented, experienced management teams have a realistic opportunity to obtain VC funding.

An early-stage company can benefit in a number of ways from being funded by a venture capitalist. In addition to providing capital, a VC firm can be a valuable advisor to its portfolio companies. Indeed, venture capitalists typically are not passive investors, and VCs that lead deals expect, to varying degrees, to be actively engaged with their portfolio companies. Thus, to reap the greatest benefit from its fundraising, an early-stage company should analyze the strengths and weaknesses of its potential VC investors.

FSA tracks VC funding for private-equity fabless semiconductor companies (Figure 13.1). The fabless semiconductor industry has been fueled by innovation and VC dollars.

No venture capitalist firm claims to be all things to all people. In other words, no VC has the professional talent, capacity, experience and available funds to be the best option for all early-stage companies seeking funding. Entrepreneurs expect VC firms to do extensive due diligence prior to making an investment. The converse should also be true: entrepreneurs should fully understand what a prospective VC investor brings to the table before accepting the investment.

Obviously, compromises must sometimes be made if the search does not uncover a willing VC firm that is a perfect fit for the entrepreneur. Nevertheless, regardless of the amount of

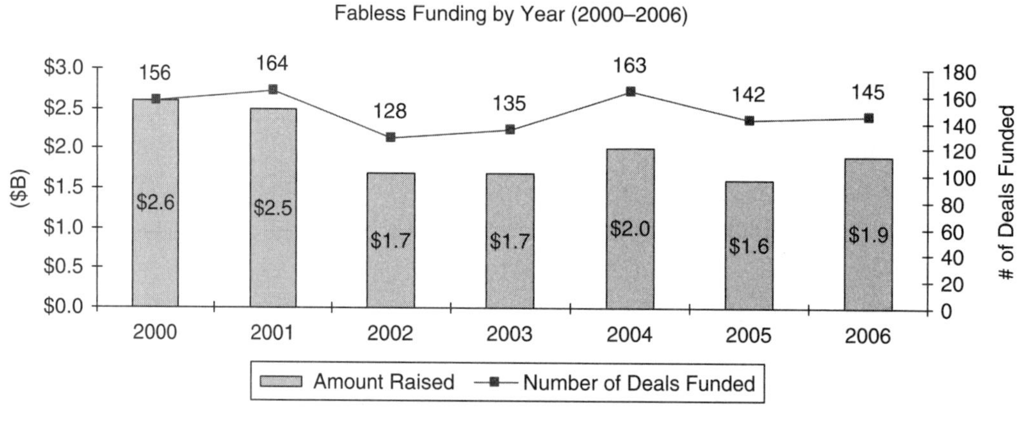

Figure 13.1

capital sought or the timing of the transaction, any early-stage company seeking funding from a venture capitalist should gather as much information as possible about the VC before accepting a term sheet. A consideration of the following factors may be useful in that process.

13.2.1 Factors to Consider

13.2.1.1 Availability of Funds

Does the VC currently have funds available to make the investment (and potential follow-on investments)? As the economy has worsened over the last few years, VCs have earmarked a greater percentage of their available funds to support their existing portfolio companies. That leaves less money available for new investments. An entrepreneur should, therefore, ask a prospective investor not only about the total dollars available for investment, but also about the percentage of the available funds expected to be needed to support the VC's existing portfolio.

Entrepreneurs should also be comfortable that a venture capitalist is willing and able to close an investment without unnecessary delays. However, the entrepreneur should be aware that the average time between the initial contact and funding is currently approximately six months. A few years ago, when the supply of funds available for investment exceeded demand and competition among VCs for the best deals was intense, the process often took 2 months or less. Today, start-up companies that expect to burn through their available cash within 60 days may be better served by seeking financing elsewhere.

13.2.1.2 Relevant Experience

Most early-stage companies look to their lead investors for advice, through the investor's representative on the board of directors or on a less formal basis. The quality of the advice is affected dramatically by the extent of the VC's relevant knowledge and experience. TL Venture's Guy Hoffman said that his firm will make an investment only if they feel that what they can offer will "contribute to the company having an unfair advantage over its competitors." According to Hoffman, "The most important factor is track record – has the VC invested in successful companies at a similar stage of development?"

Another factor that should be considered is whether professionals within the firm have experience with, and therefore understand, the company's business, industry and market. In Hoffman's view, this domain expertise, while significant, is somewhat less important than the VC's track record with similar-stage companies. A firm's industry focus can easily be determined by reviewing a list of its prior investments, which usually can be found on the firm's Web site. Ideally, of course, an entrepreneur will have determined whether his or her company fits a VC's investment profile before making the initial contact with the firm.

A VC's general business and transactional experience may be useful to the entrepreneur in a variety of circumstances. For example, a venture capitalist with extensive marketing experience may be extremely valuable to a start-up company in performing crucial market analysis and developing a solid marketing strategy. Similarly, an entrepreneur who does not have a strong financial background may benefit greatly from the venture capitalist's experience with complex financial modeling and analysis. A VC may also be able to assist in negotiations with third parties such as strategic partners or vendors.

A technology start-up may also want to determine whether a VC has professionals on staff who have technical backgrounds that will allow them to have more than just a layperson's understanding of the company's products or services. This factor, however, is probably secondary to the factors described above. A VC firm with a relevant track record and domain expertise can engage its own advisors as needed to develop the necessary technical knowledge and understanding.

13.2.1.3 Relationships

Does the VC have other relationships that may benefit the company? Does the VC's portfolio include, for example, investments in companies that may become customers, vendors or business partners of the company? If necessary, can the venture capitalist introduce the company to qualified attorneys, accountants or other advisors? The entrepreneur should determine what the experience of the firm's existing portfolio companies has been in this respect. A venture capitalist with an impressive Rolodex will benefit the company only if he or she is willing to use it on behalf of the firm's portfolio companies.

13.2.1.4 Capacity

Does the VC have the bandwidth and the motivation to serve as an effective strategic partner? What is the firm's ratio of professionals to portfolio companies? More specifically, what are the key individual's other commitments, and might they interfere with his or her ability to help the company? The entrepreneur should be comfortable that the individual professionals at the VC firm, particularly the individual who will serve on the company's board of directors, will be able and willing to devote the time and energy necessary to be a valuable asset to the company. The entrepreneur should discuss these issues with the VC. He or she should also contact the VC's other portfolio companies to discuss the extent of their post-investment contact with the venture capitalist and their general impressions about the firm's willingness to be an active partner rather than a passive investor.

The entrepreneur may also want to discuss with other portfolio companies the extent to which professionals within the VC firm, other than the company's primary contact, have become involved. Professionals within a VC firm may have complementary skill sets and relationships. The extent to which VCs are willing to devote the time and effort necessary to assist the firm's portfolio companies, other than the ones on whose boards they serve, varies among firms.

13.2.1.5 Assistance with Recruiting

Closely related to the relationships and capacity issues is the extent to which the VC can be expected to participate in recruiting talent for the portfolio company. While a venture capitalist can be an extremely valuable board member and strategically, the company's success will ultimately depend on the quality of its own management team. For many companies, one of the primary benefits that a VC firm can provide is helping locate and recruit talented management and technical personnel. The willingness to perform this function varies among firms, as the recruiting process can take significant time and effort.

13.2.1.6 Ability to Attract Other Investors

Some VC firms are frequently the lead investors in financing transactions, while others prefer to be co-investors. Among those that lead transactions, some firms have better track records than others with respect to attracting co-investors and investors in follow-on rounds of financing (i.e., their investments are seen as "smart money" by the VC community). Conversely, the participation of a "non-strategic" investor in an early financing round may discourage some VC firms from participating in later rounds.

13.2.1.7 Staying Power

Although it is tempting to focus only on the amount and terms of the current financing transaction, it is important for an entrepreneur to understand the VC firm's track record of making subsequent investments in its portfolio companies. The unwillingness of a first-round investor to make a follow-on investment may be a kiss of death for the entrepreneur's ability to raise funds from other VC firms in later rounds. In part, this is a question of availability of funds. Does the VC have adequate funds to provide follow-on support to all its portfolio companies and preserve its historical ratio of total investment to first-round investment?

Hoffman offers one caveat when evaluating staying power: "For any specific company, a VC may have a good reason – such as a significant disagreement with management on philosophy – for not making a follow-on investment. An entrepreneur should look at the VC's entire portfolio to discern the firm's record of making follow-on investments."

In the current environment, it may also be useful to determine whether a VC firm has made follow-on investments at a price above the first-round price, given the stigma attached to a follow-on investment made at a lower price, a so-called "down round." Again, however, there may be legitimate reasons for a down round in individual cases.

The willingness to make follow-on investments is not the only measure of staying power. An entrepreneur should discuss with a venture capitalist's existing portfolio companies whether the VC has remained an active advisor after the closing, and whether the firm has remained actively engaged with the company in bad times as well as good.

13.2.1.8 The Process

Entrepreneurs may find that gathering information on VCs is not as difficult as they might expect. Surprisingly, however, many entrepreneurs do not take advantage of the opportunities for learning about their prospective investors. "In my experience," said Hoffman, "nine out of 10 entrepreneurs don't research VCs at all. That's amazing to me, given how readily available the information is." In fact, Hoffman sometimes questions whether a start-up company that has not done its research on Hoffman's firm can be relied upon to do adequate due diligence on its prospective customers, employees or other business partners.

The best starting point for finding information about a VC firm is the firm's own Web site. Most VC Web sites contain a wealth of information on the firm's existing portfolio, past portfolio, investment philosophy, investment criteria and the qualifications and experience of the firm's professionals. They also typically contain links to Web sites of the VC's portfolio companies. Using these sources in combination, an entrepreneur will generally be able to determine the venture capitalist's record of making follow-on investments; other firms that have co-invested with the VC; the typical size of the firm's initial and follow-on investments; and other valuable objective and measurable information.

Some of the more qualitative factors are not as easily evaluated solely on the basis of the data included on these Web sites. The entrepreneur should contact several of the VC's portfolio companies to discuss these issues. Obviously, some portfolio companies will be more willing than others to engage in detailed conversations. Regardless of the extent of the conversation, these conversations will, at the very least, reveal whether the company has had a positive experience with the VC.

When performing due diligence on a VC firm, the entrepreneur should focus on the individual professional within the firm who will be the company's primary contact. Although, ideally, the entrepreneur will be able to draw on the talents of a number of people within the firm, the company's relationship with its primary contact will be the principal determinant of the value of the relationship. For this reason, another portfolio company's experience with one professional within a VC firm may not be particularly relevant to an entrepreneur whose relationship will be with a different professional within the same firm.

Finally, one of the best ways to obtain information about a VC firm is simply to ask. "Most VCs I know are straight shooters," said Hoffman. "You would be surprised at how frank they can be about their views on a company's business plan and prospects and about their own firms, particularly after they have developed a rapport with the entrepreneur."

In fact, the ability to develop a rapport with a venture capitalist fairly early in the process is itself an important consideration in finding the right VC. Open and honest communication between the entrepreneur and the VC is the key to a successful, long-term relationship.

PART 4

The Fabless Business Model: A Look into the Future

CHAPTER 14

Perspectives into the Future of Fabless

14.1 Keeping Up with the Pace of Change in a Fabless World

The market dynamics faced by fabless semiconductor companies today are forcing them to re-evaluate and adjust their business strategies. In many cases, the success of the past cannot be relied on to ensure success in the future. Fabless firms must know the evolving manufacturing processes and business models, always be aware of their customers' changing expectations and they must have an intimate understanding of their own capabilities.

14.1.1 Looking Ahead at Manufacturing Processes

14.1.1.1 Merged DRAM and Logic

Silicon processes that merge DRAM and logic are commercially available, laying the groundwork for the market evolution to systems-on-a-chip (SOC). When designing SOCs, system architects are no longer constrained by data-transfer bottlenecks between logic and memory at the board level. The removal of these logic-memory bottlenecks represents a significant opportunity for performance improvement in many applications. Fabless firms that exploit these new architectures to improve the performance and functionality of their products will be able to maximize their share of the SOC market, estimated to be a multi-billion dollar industry in 3 years.

14.1.1.2 The Proliferation of Semiconductor IP

Fifteen years ago, most vertically integrated electronics companies outsourced printed circuit board (PCB) manufacturing. Now they are also completing the outsourcing of PCB assembly and test. The industry has evolved to the point where the most valuable item on a PCB has become the proprietary chip, the board itself being the necessary vehicle to connect the chip to the "system". The emergence of the "fabless chipless" IC company, and the proliferation of reusable semiconductor IP is now enabling IC designers to focus on the portions of the chip design that are truly unique and provide the highest value to their customers. For fabless semiconductor companies, such as peripheral IP (I/O cores, ADCs/DACs, PLLs, etc.) and even embedded processors to become readily available, the valuable item is becoming the proprietary IP deep inside the chip. Fabless firms must now ask themselves, "What is it that we are really selling to our customers?" "Should we be threatened by the proliferation of cores, or do we have proprietary IP that adds value?"

14.1.2 More SOCs

ASIC flows made custom ICs available to the masses. Now IC designers need to incorporate the following into their designs as the market moves to an SOC focus:

- Complex semiconductor IP from various disparate sources – some hard and some soft
- Full-custom DRAM
- Mixed-signal cores
- Proprietary blocks.

This evolution to SOC can be very intimidating to the most seasoned IC designer, not to mention system designers. Design service providers who are able to help customers navigate these waters are viewed as strategic, high value-added partners.

14.1.3 Need for Flexibility

In today's evolving design services market, customers are demanding flexible business models from their design partners. In fact, the number of business models a design firm must offer are almost as numerous as the number of customers that they work with. The following are just a few business models used in the industry:

- *Free IP model*: This model provides customers with design services and IP at an extremely low or non-existent cost. The benefit to the customer is little up-front investment to develop a new product. Design companies benefit by back-end loading such a partnership with wafer surcharges, royalties or a commitment to purchase design tools or other services. In the end, both win, provided the designed program goes to production at the forecast levels. The downside occurs if the program fails to move into production or the anticipated volume never materializes.
- *IP licensing*: This model, proving to be quite successful, usually involves high-end IP such as RISC processors, controllers, DSPs, etc. IP in this area is extremely strategic in the development of a system-solution and is readily available to those who have the available funds to purchase the license. Examples include ARM, ARC, MIPS and Rambus.
- *Risk/Upside Sharing*: A firm that has done sufficient market research and possesses extensive applications know-how, can adopt a turn-key design services model in

which some of the risk and much of the upside is shared with its client. For example, the design services provider may collect only minimal service revenues prior to the chip going into production, but significant downstream royalties. This model often looks more like a partnership, with the design services client providing a sales/support channel and system-level components such as an integrated software application.

- *Go it Alone*: This is the true fabless semiconductor company model. The design company assumes full risk. The market research, sales channel and customer development, product design, foundry and back-end partnerships are all arranged by the company. Given the right market dynamics, a fabless firm can be hugely successful selling its own branded, high-value merchant silicon products. Accurate estimation of investment funding and risk is critical for this model, not to mention the successful execution of the business plan. However, the benefits are plentiful.

14.1.4 Climbing the "Value Curve"

For fabless semiconductor companies offering design services, and even for those offering merchant silicon in some cases, the fiercest competition can come from the customer's internal ASIC design team. It is critical to be viewed by the customer as a partner or a strategic supplier and not as a "body shop" of IC designers. The higher the value of the product provided to its customers, the more the fabless firm is viewed as a strategic partner in their success. This progression is depicted in Figure 14.1.

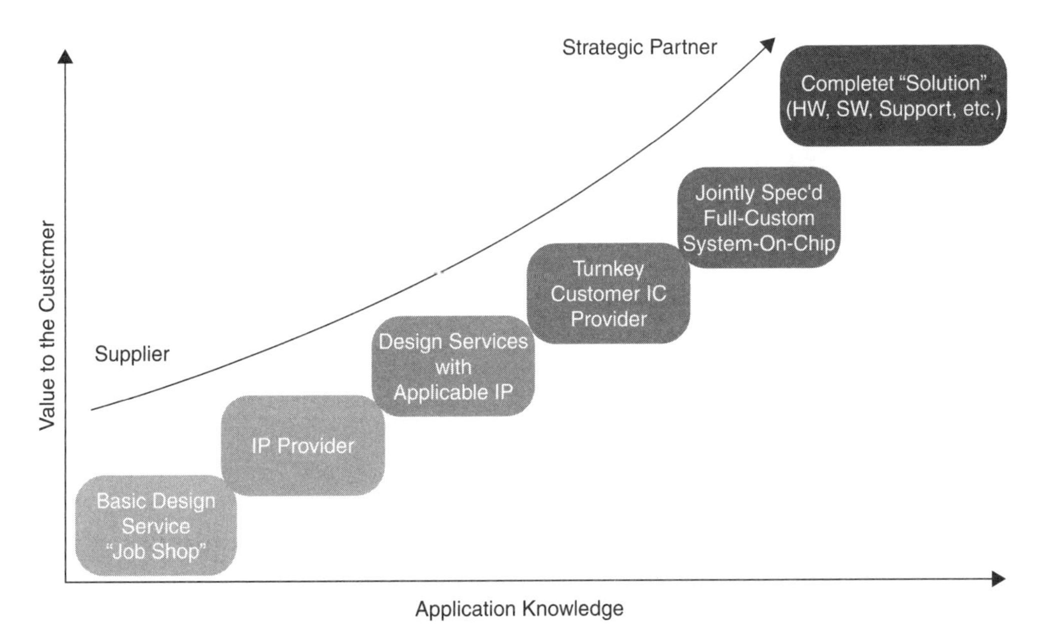

Figure 14.1 Design Service Providers Value Curve

14.1.5 Do Not Rest on Your Laurels

Consider the convergence of DRAM with logic. This new IC manufacturing capability ensures that the future will continually evolve with emerging systems applications that combine memory and logic effectively. Currently, DRAM design is still a very specialized field. DRAM designers do not enjoy the rich suite of EDA tools that have made life easier for ASIC developers in recent years. Nevertheless, the merging of logic and memory manufacturing processes will require tool vendors to augment their tools for simulation, layout, floor planning, extraction and annotation to include DRAM elements. It will be a slow evolution at first, but it will happen.

14.1.6 Focus, Focus, Focus...

Fabless firms need to deliberately choose markets in which their SOC capabilities set them apart from the rest of the playing field. The availability of merged DRAM/logic processes, and the proliferation of semiconductor IP, has opened up a world of opportunity that the semiconductor industry has not seen for some time. The winners will be those companies that are quick to seize opportunities and are best able to match market/customer demands to their core area of expertise. Concentrating on market opportunities that best fit the company will be the key. Companies that focus on areas in which they have the best applications knowledge, core competencies, the right kind of dynamics in terms of customer concentration, price sensitivity and growth will improve the likelihood of a successful outcome and a share of these newly created opportunities.

14.1.7 Conclusion

The dynamics in today's semiconductor industry are making it necessary for fabless firms to review what they truly bring to their customers. There are many possible roads to success, but they may be different from the ones previously followed. Fabless companies that are aware of the changes afoot and address them in a focused and deliberate manner will be the most successful.

14.2 Foundry Roadmaps: Partnering, Leading and Innovating

The evolution of process technology roadmaps for foundries has paralleled the course of fab-lite and fabless semiconductor sourcing strategies. Traditional foundry roadmaps were based upon the outsourcing of standard CMOS manufacturing for high-volume logic products. This was a successful formula for the growth of the foundry industry because the technology requirements were pre-defined by the customer. A focus on implementation of processes resulted in efficient foundry operations that, over time, set new standards for manufacturing excellence.

Support of multiple variants of a base technology, such as 0.25-micron digital CMOS with three-, four- and five-layer metal options, came with the growing number of fabless and fab-lite companies in more diversified end markets driving the growth of foundry business. Mixed-signal (MS) ICs grew with the integration of analog and logic functions, leading to the addition of support for analog features, such as isolated resistors and linear capacitors, to supply integrated device manufacturer (IDMs) that were rapidly outgrowing their own specialized manufacturing capacity.

The sheer volume of ICs consumed by wireless communications devices initiated a path toward RF CMOS to serve fabless customers who saw an opportunity in areas dominated by specialty technology products developed and manufactured by IDMs that were able to optimize their processes for high-volume applications. This step was a key one along the path of foundry roadmap development since it marked a transition from serving the needs of customers with technologies already available at IDMs, to establishing new technology variants that enabled fabless companies to enter product areas that were previously the domain of IDMs (Figure 14.2).

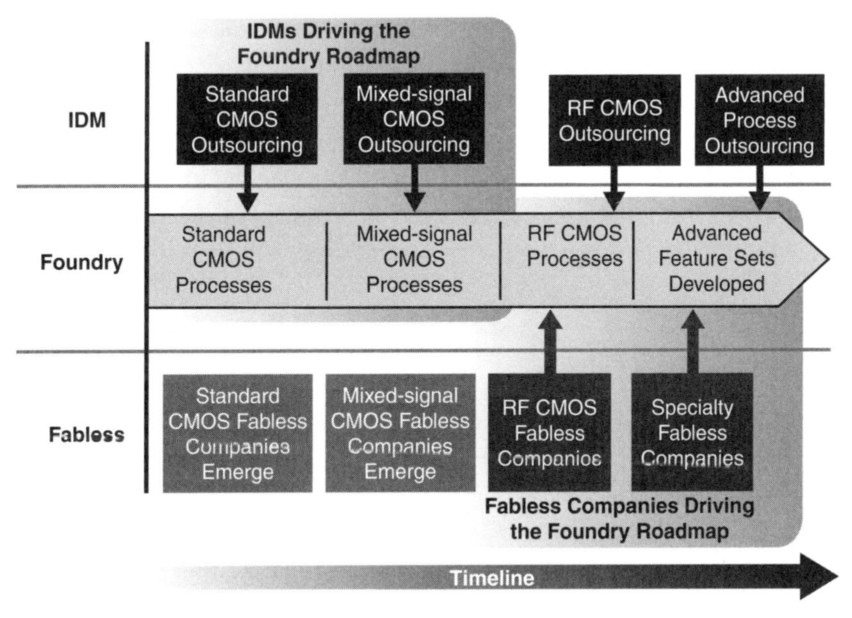

Figure 14.2 Driving the Foundry Roadmap and the Increasing Role of the Fabless Model

Other examples exist where foundries lead the way based on their customer's input, such as the integration of Flash or DRAM with logic, or the ability to deliver leading-edge silicon germanium (SiGe) BiCMOS for high-performance applications. The evolution of foundry roadmaps has moved from serving as a manufacturing partner to its current state, where foundry-based technology challenges the best in the semiconductor industry, meeting the technology leadership requirements that enable the product leadership of many fabless and fab-lite companies.

14.2.1 Roadmaps are More than Process Technology

It is not enough to just deliver process technology in an environment where companies depend on their foundry supplier to keep them competitive against IDMs. Providing access to advanced technology and enabling time-to-market advantages are key in determining end-market success. Since the foundry roadmap has moved beyond just meeting the requirements of second sourcing and has become a critical source for new and differentiated technologies, delivery of a complete ecosystem of processes, models, design kits and circuit IP has become an essential component of the foundry roadmap. Without the delivery of this additional layer of capability, a large number of companies would have to create their own design infrastructure, and their ability to do so would be a key differentiator. In more cases today, the key differentiator is the applications knowledge and the IC architecture or design that best meets the needs of the system product. An ability to deliver representative models, design kits and simulation capability has become expected of the foundry in cooperation with EDA tool vendors who deliver the underlying software tools, while the foundry supplies the additional layer that ties the process capability to the anticipated silicon output. Delivery of circuit IP that forms the building blocks for custom IC designs is expected from multiple sources, but management of the available IP portfolio falls back to the foundry to communicate a consistent view of the validation level to the user.

14.2.2 More than One Roadmap

One of the current challenges for foundry suppliers is the diversity of technology requirements that are presented by customers serving a large variety of end markets. At the same time, economies of scale demand that a large number of processes be based on the same fab equipment and capability for each generation of process, generally defined by the minimum line width for that node. Within the same factory, the foundry is often expected to deliver high performance, low power, embedded memory and RF capability; sometimes this is expected on a single wafer. To be successful, a foundry must continue to expand the feature set available to customers by developing and integrating key process modules that provide a path for integration of multiple functions in an IC, or multiple ICs within a system, while leveraging similar capacity across multiple technology variants. The foundry roadmap is defined by its customers' pursuit of higher performance, lower power and other capabilities, which ultimately result in reduced cost per function.

14.2.3 High-Performance Graphics Leading the Way

The largest graphics and programmable gate array companies are fabless, making them the most significant drivers of new foundry technology where computational speed and routing density are the primary considerations. Rapid adoption of 130-nanometer and 90-nanometer technology nodes has provided benefits in terms of speed and integration level. Leading logic foundries moved quickly to master the implementation of low K dielectrics to reduce interconnect parasitics that would otherwise have limited performance. Embedded high-density memory has improved access times and allowed development of more efficient SOC architectures.

14.2.4 Low Power Fueled by Mobility

Fueled by the growth in the number of portable electronics devices, support has grown for process variants that provide low operating voltage and low stand-by current at the highest available level of integration. Applications, such as the mobile phone, benefit from the lower cost per function made possible by including many digital functions on a single chip. Low-power libraries and embedded memories optimized for power rather than speed are a necessary part of the foundry roadmap to support a low-power design strategy.

14.2.5 MS/RF for Communications

What started as a minor variant in the foundry roadmap to support the integration of analog circuits with digital functions has evolved into an increasingly important part of the roadmap to address higher levels of SOC implementation, as well as stand-alone RFICs. While generic CMOS process technology is often used for the design of MS SOCs, leading companies have made incremental features such as triple-well isolation, low threshold devices, linear metal-insulator-metal (MIM) capacitors and thick inductor metal standard components of an RF CMOS roadmap. Focus on an analog mixed-signal (AMS) design platform to support foundry customers includes delivery of device generators, RF models, extraction of parasitics and accurate simulation capability. The fast pace of change in the evolution of the end-market applications for RFICs has been supported by advanced combinations of features, such as embedded Flash in low-power RF CMOS, that demonstrate the breadth of process capability available today from foundries.

14.2.6 Specialty for Integrated Analog and RF

There are a large number of application-specific silicon technologies that have continued to be primarily sourced from IDMs, but have now made the transition to being supported by foundry roadmaps. These specialty processes have unique attributes that enable their use in more highly integrated MS ICs where the analog device capability dominates the performance requirements. One of the largest requirements is for display driver ICs that support up to 40 volt interfaces with 0.18-micron CMOS. Another major area supported by foundry roadmaps is advanced analog and RF, where several generations of high-speed SiGe BiCMOS are providing solutions for a growing number of RF transceiver, RF power amplifier, TV tuner and emerging high-performance applications. An overall trend toward integration of analog capability has determined the direction of specialty foundry roadmaps that support increasingly diverse process capability with modular technology, AMS design platforms and application-specific circuit IP.

14.2.7 MEMS: Coming Soon?

With no clear established foundry model for microelectromechanical systems (MEMS), it is premature to discuss the foundry roadmap for this technology. However, the emergence of fabless MEMS start-ups is an indicator of an opportunity in the industry for foundries to partner with their customers to establish new technology capability and roadmaps. MEMS are indeed growing in significance and have a role in this industry. It is this interdependence and level of cooperation between customer and supplier that will result in continued breadth and depth of technology and associated support that the foundry will be able to offer.

14.2.8 International Technology Roadmap for Semiconductors

The International Technology Roadmap for Semiconductors (ITRS) is considered the benchmark for semiconductor technology roadmaps. It is also recognized as a mechanism for chipmakers, equipment and materials suppliers, and researchers to coordinate their understanding of requirements for the entire semiconductor device supply chain to move forward, meeting the challenges in the near and long term. As a benchmark, it is seen as the timeline to beat by the most aggressive chipmakers, including foundries that routinely deliver process technology in advance of the ITRS timeline, but it is nevertheless indicative of the overall trends in the industry. The diversity in the technology base now available in foundry roadmaps mirrors the broadening of relevant processes, devices and structures that have been highlighted in the most recent ITRS updates.

Roadmaps serve by combining key elements of different roadmaps, adding specialized capability to standard processes, or guiding the most important attributes of advanced linewidth technologies (Figure 14.3).

Similarly, there is significant momentum behind development of the design tools required for integration of large digital systems in SOC products, and the needs of the analog/RF designer are beginning to be addressed with a similar focus on tool functionality and integration. However, the expectations of IC designers continue to stretch the capability of available design tools. An industry view of priorities for design tool enhancements would benefit both the tool developers, in terms of efficient use of R&D spending and more tool sales, and

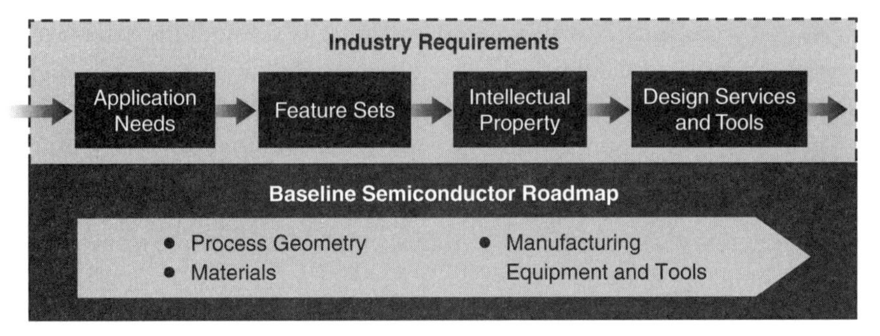

Figure 14.3 Defining Technology Roadmaps to Meet Changing Industry Requirements

users who get the tools they need to speed products to market. The concerted effort already underway to create a more efficient business model for the adoption of circuit IP from a variety of third-party providers could be the first step toward establishing IP roadmaps that better fill the needs of the design community and foster growth of this segment that is critical to the fabless semiconductor ecosystem.

14.2.9 Roadmap to the Future

The pace of change makes it more difficult to predict the look of the semiconductor technology landscape within the next decade. Advanced linewidth CMOS technology has been adopted earlier than anticipated, often with the use of equipment and materials proven in the previous generation rather than with exotic new materials. In addition to new foundry players who have demonstrated the maturity of advanced silicon manufacturing by ramping up quickly, a new set of focused specialty foundries has emerged that is able to provide application-specific technology that was historically difficult for the fabless community to access. Today, there are predictions that transistors will be re-invented to continue the scaling of logic and memory. Regardless of the changes along the way, the relationship between foundries and their fabless and fab-lite customers will continue to strengthen, based on the growth of the foundry business model and their combined ability to navigate the most effective roadmap to the future.

14.3 Semiconductor Manufacturing in the 21st Century

The market dynamics and the capital investment required for semiconductor manufacturing in the 300-mm era will have far-reaching effects on the business models of IDMs, fabless semiconductor companies and foundries. The effects will be even more dramatic when the industry begins the era of 450-mm manufacturing.

In the 300-mm era, consolidation will accelerate, primarily through the extensive utilization of joint ventures by majority of the top 10 IDMs. Although the top three IDMs might participate in joint ventures as a means of satisfying peak production demand, the primary participants will be smaller IDMs, fabless semiconductor companies, and foundries such as TSMC, UMC, Chartered and SMIC.

In addition, the reliability of Moore's Law as an accurate predictor of shrink effects in chip technology will come into question – not because of technical barriers in IC manufacturing but because of economic barriers. Ongoing shrinks are also hampered back-end inefficiencies and rising costs for non-silicon materials.

14.3.1 Semiconductor Market Share

As the industry moves aggressively into the 300-mm era, it will be extremely difficult for any IDM with less than 5 percent market share to compete successfully without pooling its resources – and its utilization of wafers.

14.3.2 The Evolving Manufacturing Landscape

The most obvious reason for consolidating manufacturing resources is the cost of constructing a 300-mm fab, now estimated at about $4 billion. Less often noted in the popular media but no less important to corporate profitability is the manufacturing volume of a 300-mm fab – and more specifically how to keep the fab operating as near to 100 percent loading as possible, which is close to being a prerequisite for profitable operation. Today this is already a challenge, even for the largest IDMs and the largest foundries.

The semiconductor manufacturing landscape can be represented conceptually by the illustration in Figure 14.4. The boxes on the left side of the diagram that diminish in size as they go down represent the IDMs and fabless semiconductor companies ranked by revenue. The geometric shapes on the right side represent the proportionate amount of manufacturing that supports the IDMs and fabless companies.

Comparing the left and right sides of the diagrams, one can see how most the manufacturing requirements of the large OEMs are satisfied with captive or in-house fabs. The medium-sized IDMs satisfy their production demands with a combination of their own fabs, joint venture fabs and foundries. At the bottom of the diagram, fabless semiconductor companies rely primarily on foundries. In every instance – even for the largest IDMs – foundries play some role in satisfying wafer demand.

In the 300-mm era, consolidation through joint ventures will change this conceptual landscape radically for two reasons. First, the cost of 300-mm fabs, now roughly $4 billion, limits the number of companies that can afford it. Secondly, it is difficult to keep them full and therefore profitable as most IDMs do not have sufficient internal chip demand to load a 300-mm fab.

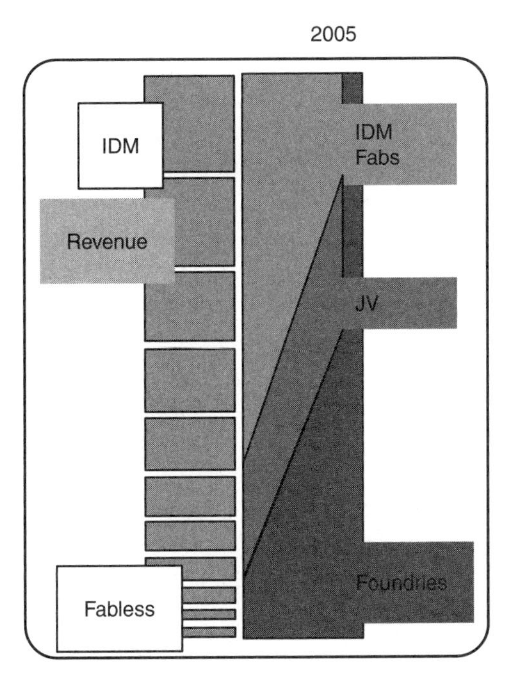

Figure 14.4 Foundries and IDM fabs dominated IC output

The smallest change in the relationship between demand and type of supply between 2005 and 2010 will be for the top two or three IDMs. They can afford 300-mm fabs because they can keep them full.

The largest change will be the relative proportion of manufacturing that will take place at joint venture fabs, which is increasing, at the expense of the small portion that is being produced in captive IDM facilities. This development will not erode the foundry segment, however, because the foundries will probably be active participants in the joint ventures.

Operating the joint venture fab at full or close-to-full capacity from its inception is made possible by introducing a new technology in a three-phase process. In the embryonic phase, a pilot fab is used to develop the technology and do some pilot production. During this period, the joint venture that will ultimately build a fab is formed and fab construction begins.

As soon as the technology is proven, the growth phase is initiated with production capacity being supplied by a foundry. About midway through the growth phase, the joint venture fab comes on line. Any additional manufacturing demand is shifted to the joint venture fab. Assuming a typical demand curve, the fab's manufacturing ramp is very steep.

In the technology's mature phase, the wafer demand is shifted from the foundry to the joint venture fab, including some of the foundry's original capacity. Using this strategy, the joint venture fab is virtually assured of being full – and profitable from its inception. Since it avoids the lower end of the demand ramp, it can be highly profitable.

With two or three partners participating in the joint venture, each partner can reap the benefits of a 300-mm fab at far less capital investment than building the fab alone, as well as the benefits of an average cost per wafer based on running the fab at full capacity.

The impact on the industry is shown in Figure 14.5, which illustrates a conceptual landscape of the semiconductor manufacturing industry in 2010.

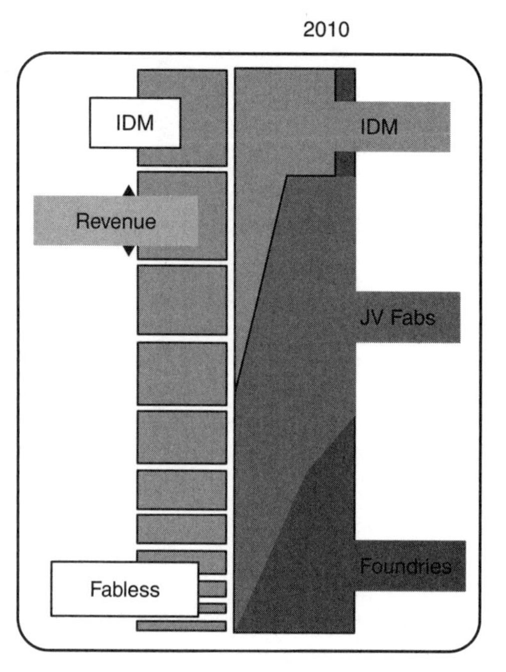

Figure 14.5 By 2010, joint venture fabs will capture the largest portion of IC manufacturing

Figure 14.5 shows a much larger share of worldwide manufacturing occurring in joint venture fabs between medium-sized IDMs, foundries and fabless semiconductor companies. Manufacturing in fabs owned 100 percent by foundries is proportionally smaller.

But it should be kept in mind that 20 new 300-mm fabs are now on line compared to 2005 so the total volume in 100 percent owned foundries may not drop. Finally, the top three IDMs will still do the lion's share of their manufacturing with peak demand going to the foundries.

In an interview with Wim Roelandts, president, chairman and CEO of Xilinx Inc. (and former chairman of FSA) in early 2007, Roelandts predicted the industry would only see three chip models in the future: fabless/foundry, memory and Intel. Much has already been stated about the fabless model. Memory makers will continue to build their own plants, though the industry is witnessing partnerships being formed in this segment as well. And Intel will continue to build its own fabs that are tuned for its processor lines, but even Intel is taming advantage of the outsourcing strategy to a lesser extent.

14.3.3 Industry Outlook

The semiconductor industry ended fairly strong in 2006, recording sales of IC at $209 billion, 8.5 percent above 2005's figures (Figure 14.6).

WW Electronic Sales ($B)	2006	2007 (e)	YoY Change
WW Chip Sales	1453.5	1538.3	5.8%
IC Sales ($B)	209.1	220.1	5.2%
IC Units (B)	138.2	152.9	10.7%
ASP ($)	1.51	1.44	−4.9%
WW Chip & Display Equipment			
Total Sales ($B)	61.4	64.3	4.8%
IC Equipment ($B)	53.5	55.8	4.2%
Wager Fab ($B)	29.5	30.9	4.7%
Test & Related ($B)	12.2	12.6	3.9%
Assembly ($B)	3.5	3.3	−6.3%
Service & Support ($B)	8.3	8.9	7.5%
Display Equipment	7.8	8.5	8.4%
WW Capacity Utilization			
Chip Production (MSI)	5885.3	6351.8	7.9%
Chip Capacity (MSI)	6365.4	6965.9	9.4%
Front-End Utilization (%)	92.5%	91.2%	
Test Utilization (%)	93.9%	91.5%	
Assembly Utilization (%)	90.1%	90.5%	
Source: VLSI, January 2007			

Figure 14.6 Worldwide Chip Sale sand Related Manufacturing Rates

The Foundry market forecast is listed in Figure 14.7.

	2005	2006	2007	2008	2009	2010	CAGR (%) 2005–2010
Foundry Revenue	$18,423M	$21,885M	$25,280M	$32,485M	$29,853M	$30,689M	10.7%
Growth (%)	−2.2%	18.8%	5.5%	28.5%	−8.1%	2.8%	-
Source: Gartner Dataquest, August 2006							

Figure 14.7 Foundry Market Forecast (2007)

14.3.4 The 450-mm Era

The market dynamics and manufacturing paradigm that will drive the 300-mm fab era are evident enough for credible forecasting. The 450-mm era, on the other hand, presents a less predictable forecasting environment.

According to a spokesperson from Synopsys Inc., only an elite group of companies will be able to afford to develop leading-edge IC designs over time. At 45-nanometer:

- A new 300-mm fab costs approximately $3 to 4 billion, depending on your source.
- Process technology costs about $2.4 billion (includes CD tools, process models and related costs, but not the fab equipment).
- A "mask set" may run as high as $9 million or more.
- IC design costs range between $20 million and $50 million.

Technical issues that affect success at these nodes include:

- Process variation at the 45-nanometer node, which is causing significant pressure on preventing chip failures due to variation issues.
- Power and leakage issues have yet to be resolved.
- New physical effects in lithography remain unchallenged in large part.
- Product-specific yield learning is still new territory.

At 32-nanometer, data is less clear, but estimates forecast:

- A new 300-mm fab could cost upwards of $10 billion.
- Process R&D costs could reach $3 billion (includes CD tools, process models and related costs, but not the fab equipment).
- Design costs could hit $75 million.

Neither the IDMs, fabless or foundries will remain unscathed. Plant costs will escalate faster than expected due to the migration to smaller technology nodes, advanced semiconductor manufacturing, handling and test equipment costs will skyrocket, as fewer are sold, but the investment in R&D remains extraordinarily high and new materials will add costs unseen in the past.

Few will be able to achieve ROI on R&D, IDM or otherwise. In Synopsys' example, to achieve ROI on a 65-nanometer node, a company must spend $1.5 billion to develop the process technology and that figure is forecast to escalate to $2.4 billion for 45-nanometer, and $3 billion for 32-nanometer.

For example, to achieve ROI for R&D on 65-nanometer, Synopsys argues a company must generate a total of $8.3 billion in sales annually. According to IC Insights Inc., in 2006, only five IDMS had revenues of $8.3 billion or more:

- Intel
- Samsung
- Texas Instruments
- STMicroelectronics
- Toshiba

And at 45-nanometer and 32-nanometer, a company must generate $13.3 billion and $16.7 billion in annual sales, respectively, to offset R&D investment for ROI. Based on 2006 figures, only Intel and Samsung could achieve this. As a result, semiconductor manufacturers must produce in large volumes for applications in the consumer electronics markets, for example. With the emerging triple play – voice, data and video – along with the added feature of mobility, termed the "quad" play, digital convergence may open up volume opportunities not seen in the past.

There are some basic calculations, however, that may at least elucidate the risks and challenges involved in the transition from 300 to 450-mm fabs. The first is the drastic decrease in the number of fabs. Today, there are about 200–250 200-mm fabs in the world. By the end of this decade, there will between 50 and 60 300-mm fabs.

Based on the extrapolation of the revenue from a 300-mm fab, each 450-mm fab that goes online would be likely to generate $15 billion per year in annual revenue if it is run at or near capacity.

This raises the question of how the additional $100 billion in semiconductor revenue between 2010 and 2013 will be satisfied in terms of production capacity. If it is completely provided by 450-mm fabs, only six additional fabs would be required.

In addition, the same basic economics that are determining the 300-mm fab landscape would likely be true at 450-mm: first, an enormous capital expense that even fewer companies could afford by themselves; and, second, the goal of having the 450-mm megafab full or nearly full to make it profitable.

There are other troubling aspects in the transition to 450-mm: the number of high-volume products that require 250 million transistors may be too limited to provide favorable economics for 450-mm fabs.

The availability of production tools is another issue. The transition from 200 to 300-mm was disastrous for the tools industry, which lost several companies in the process and has only recently recovered. With the prospect of just a few customers and an enormous investment, the tool makers will most likely postpone the 450-mm development in favor of developing a next generation of 300-mm tools.

Faced with these realities, all but the very largest IDMs are likely to be able to afford a new fab. The number might be as low as one, or perhaps the top two. Microprocessors and memory devices are the only obvious products that are produced in volumes sufficient to keep a 450-mm fab busy.

Medium-sized IDMs presumably would have to resort to the joint venture option or adopt fabless IC technology. But the joint venture option is limited by the number of partners that could be involved and still make sharing the fab's output workable for the member companies and for the fab's profitability.

The foundry landscape is just as intimidating for the same reasons. Not many foundries can afford the investment or be optimistic about having the fab run at nearly full capacity. With these complexities in mind, Figure 14.8 illustrates one likely semiconductor production landscape in 2015.

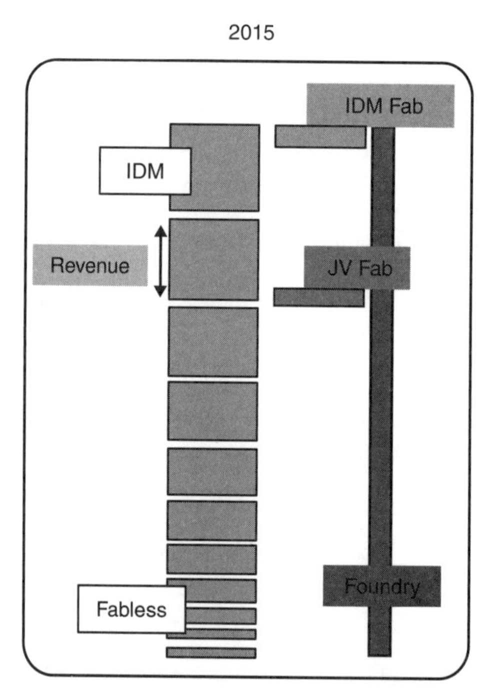

Figure 14.8 The 450-mm era will bring foundry domination to IC manufacturing

Given this analysis, the most likely scenario for the introduction of 450-mm production is that it will be delayed well past the time most forecasters project.

14.3.5 Cost: A Driving Factor in Semiconductor Industry Profitability

While cost has always been a major consideration in any semiconductor-based company given the huge investments in capital expenditures and R&D, costs and the trade-offs between time-to-market and yield have now risen to the top of the radar for all semiconductor companies. From concept, to architectural design to IP and tool investment, to production, to package and test – cost will become increasingly an issue that every phase of the organization must address.

While cost has traditionally been considered a negative concept in any business, it is actually creating opportunities for semiconductor firms. By investing in resources early and driving a strategic initiative to chip development and design reuse, companies are finding that they can save significant amounts of money in the longer term. For example:

- Design-for-test (DFT) methodologies applied early in the design phase have proven to reduce cost
- Yield optimization and DFM (design-for-manufacturability) methodologies increase yield
- Re-usable components increase productivity and decrease reinvestment
- Derivative designs with intelligent architectures allow for more diverse market applications
- Programmable and reconfigurable chips reduce time spent in the costly design cycle and reduce time-to-market.

In a presentation given in early 2007 by Wally Rhines, CEO of Mentor Graphics, Rhines commented that IDMs adopting outsourced manufacturing, process differentiation is becoming less an issue. This means that they, like their fabless counterparts, will have to rely on design differentiation focused on four key areas:

1. System architecture innovation
2. Proprietary IP blocks
3. Implementation efficiencies
4. Higher yields

By relying on system-level language and synthesizing it directly from C, dramatic improvements in speed and the reduction of logic functions can be achieved, said Rhines. Implementation efficiency and yield maximization will enable IDMs and smaller players to level the playing field.

14.3.6 Moore's Law forever?

Semiconductor die have little value to consumers until they are packaged, tested and integrated into a system. While the silicon cost per transistor, cost per megabit, and cost per MIP can reasonably be expected to decline at a rate consistent with Moore's Law for the next few years, packaging and test costs have tended to plateau and per-chip test costs are rising.

Figure 14.9 illustrates the stabilization of package cost in cents per pin over the past 10 years.

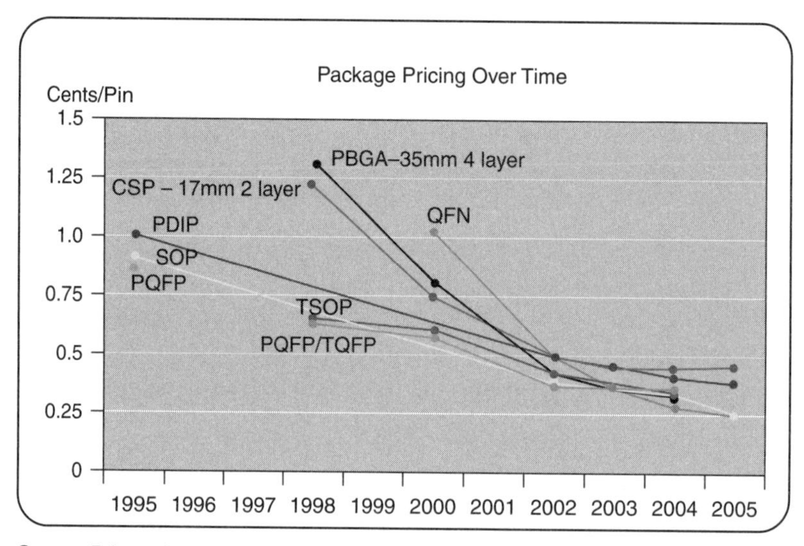

Source: Prismark

Figure 14.9 Cost per pin has not kept pace with Moore's Law

Assembly and test subcontractors have not been able to innovate rapidly enough to keep pace with Moore's Law in part because they operated in a much different environment than semiconductor companies.

Historically subcontractors have operated in a volatile market because they have been used by IDMs to handle overflow business. As a result these companies face significant diversity in their portfolios. Certainly during downturns in the chip industry their profit margins become very thin, and competition is intense. Meanwhile, both IDMs and subcontractors are also suffering from steep rises in the cost of non-silicon raw materials such as gold and copper, and the effect of increasing oil prices.

This business environment is not conducive to investments in standardization, innovation or adding capacity. As a result, there is little effective standardization among subcontractors and capacity is added in small increments. Standardization, in particular, is essential to rationalizing back-end production and spurring innovation.

Several factors will influence the ability of subcontractors to return back-end production to a cost curve more consistent with Moore's Law. A few of them include:

- Further integration, reducing the number of pins per bit
- New packaging options such as System-in-Package (SiP)
- Convergence of front-end and back-end manufacturing such as bumping
- More cooperation and coordination with IDMs
- Back-end industry consolidation

Although packaging costs appear to be controllable through standardization and innovation, the upward pressure on test costs seems to be practically irresistible. The root causes of this phenomenon are gate density and the fact that digital transistors act like analog devices.

The best strategy appears to be cost containment implemented by including DFT technologies during the design phase to achieve a higher percentage of "first-time-right" (FTR) designs despite increasing product complexity.

Philips Semiconductors has taken this approach with considerable success. Figure 14.10 shows its FTR achievements at the 120-nanometer and 90-nanometer nodes with a target of 85 percent FTR.

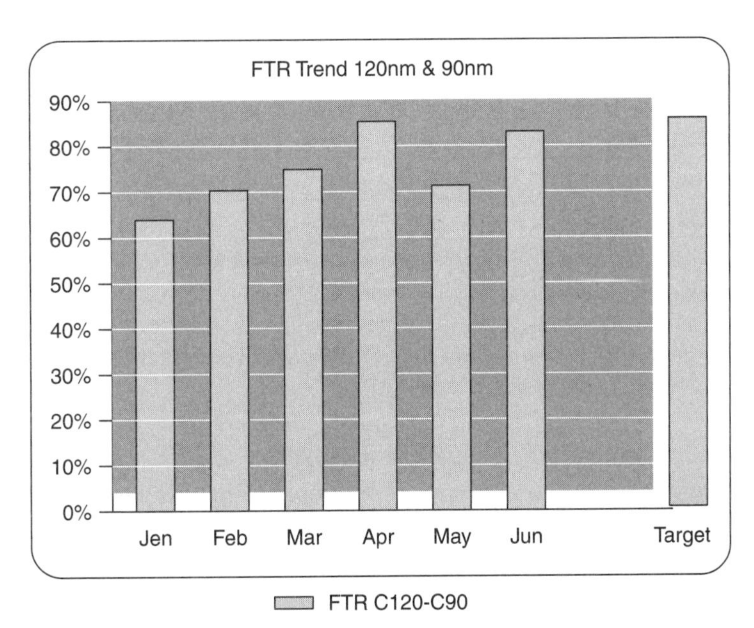

Figure 14.10 First-Time-Right goals can be achieved using DFT

14.3.7 Conclusion

Although the semiconductor production industry faces challenges, particularly in the assembly and test areas, the industry has a clear path to maintaining profitability in the 300-mm era by utilizing a consolidation strategy thought joint ventures. The forecast for the 450-mm era is not nearly as clear and is unlikely to improve in the near future. As a result, it seems likely that the 450-mm era will be delayed at the very least and perhaps not prove economically viable for some time to come.

14.4 The Emerging Dominance of China in the Technology and End Markets

Those seeking success in today's semiconductor industry should fix their gaze firmly on China. The once sleeping giant is quickly awakening with a voracious appetite for technology. In the face of seemingly insatiable demands, production in the country is at a fever pitch, and, as a result, China is now poised to overtake the United States as the major technology end market. But while the playing field may have changed, US semiconductor companies are by no means out of the game. Those that globalize quickly and are able to understand and sell into the emerging Chinese markets have the potential to thrive in the new technology world order (Figure 14.11).

TMT Category	Global Ranking	Units (MM)	2004 Growth
Mobile Phones	1	335	25%
Cable TV Subscriptions	1	128	16%
Telephone Lines	1	283	14%
Internet Users	2	94	18%
Installed PCs	4	42	17%
China leads the world in number of mobile phone users, cable households and telephone lines			
Source: Morgan Stanley Research			

Figure 14.11 China TMT Trends Remain Especially Impressive

China started to gather steam in the technology end market in 2002 and by 2005, had become a mainstream contender. A March 2006 report on global technology/media/telecom (TMT) from equity research firm Morgan Stanley identifies China as a market that has finally emerged. According to the report, although both the United States and China are now the clear leaders in the TMT market, China definitely is entering the sweet spots of technology adoption and penetration. And while it still has low penetration levels (in terms of absolute users), it is beginning to show impressive potential for more substantive 5-year growth. So if current trends continue, and it is likely that they will, China should pass the United States to become the leading end market in 2006 (Figure 14.12).

	By the End of 2004	CAGR for Next 3 Years
Volume		
Internet Users	94MM (18% YoY; <7% of population)	13%
Broadband Users	43MM (146% YoY; <3% of population)	32%
Mobile Users	335MM (24% YoY; 26% of population)	12%
ARPU		
Internet Users	$5-$6 per month	
Broadband Users	$10-$15 subscription fee per month	
Industry Revenue		
MVAS	$770MM (89% YoY)	28%
Online Gaming	$390MM (90% YoY)	37%
Online Advertising	$220MM (78% YoY)	24%
China leads the world in number of mobile subscribers		
China ranks No. 2 in Internet users with more Internet users under the age of 30 than any other country		
Growth drivers for Chinese Internet have been/are robust		
Source: CNNIC, Morgan Stanley Research		
Notes: Mobile value-added services (MVAS), average revenue per user (ARPU)		

Figure 14.12 Major Drivers for Chinese Internet Trending Up

Fueling China's emergence in the end market is its booming economy and sharp rise in discretionary income. These factors are generating huge demands for the kind of ubiquitous and mobile computing the United States has long enjoyed. For instance, China currently has 360 million cell phone users, and it is projected that there will be more than 600 million users in 5 years. China's total number of mobile users is equal to the combined total of mobile phone users in the countries that hold the three subsequent market positions. China also has huge potential for broadband penetration. So with a market of epic size ready and eager to buy every advanced technology and product it can get its hands on, China is sure to open up a huge lead over the United States in both consumption and production in the very near future.

China currently has the largest share of telephone lines (24 percent), mobile phones (21 percent) and cable television (28 percent), while the United States' share is largest in PCs (29 percent) and Internet users (22 percent). Although China is second in Internet users, it has more Internet users under 30 years old than any other country in the world. These numbers illustrate the fact that, unlike the rest of the world, China currently is mobile-centric, rather than PC-centric. This is not surprising, given the fact that the infrastructure costs of providing landline telephone service in China are much greater than the cost of mobile phone towers (Figure 14.13).

• Innvoation is coming from outside U.S. in areas related to broadband and mobile Internet
• U.S. compaines, at the margin, will likely continue to turn overseas for technological innovations
• Global investors have proven they are willing to commit rising amounts of captial to non-U.S. technology companies, particularly in Asia
• Many Asian technology companies have been criticized for being "fast followers" rather than fast innovators – but that has begun to change

Source: Morgan Stanley research

Figure 14.13 Pace of Global Innovation is Accelerating

The rush to satisfy skyrocketing demands for cell phones and other technologies has created an entirely new ecosystem of semiconductor companies in the country, and the pace of innovation is accelerating. Once criticized for being a "fast follower," rather than a "fast innovator," China now has large numbers of trained engineers and Internet and mobile phone users spurring new development. And faced with customer expectations of low-cost services, technology leaders in China are striving to meet market demands.

Shanghai-based SMIC is a good example of a successful new Chinese Internet and infrastructure company that supports the rollout of TMT. SMIC, which went public in 2005, is one of the country's most advanced facilities for logic, MS/RF, high-voltage circuit and SOC technologies, among others.

SMIC and scores of other companies like it are developing products in China not only for the Chinese market, but also for markets around the world. US semiconductor companies hoping to secure a slice of that lucrative end-market pie must position themselves in the Chinese market as part of the food chain for meeting the growing demands (Figure 14.14).

The Chinese end market comprises the complete spectrum of the industry, from medical technologies that improve healthcare to Internet communication, wireless, infrastructure and communications technologies. And one thing all of these technologies have in common is the need for semiconductors.

Although the dominant market has moved to the Far East, US semiconductor companies can certainly play, and have been playing, in that market. The challenge for companies seeking

Revenue Composition – Top 13 China Internet Companies CQ3 Annualized – S2B

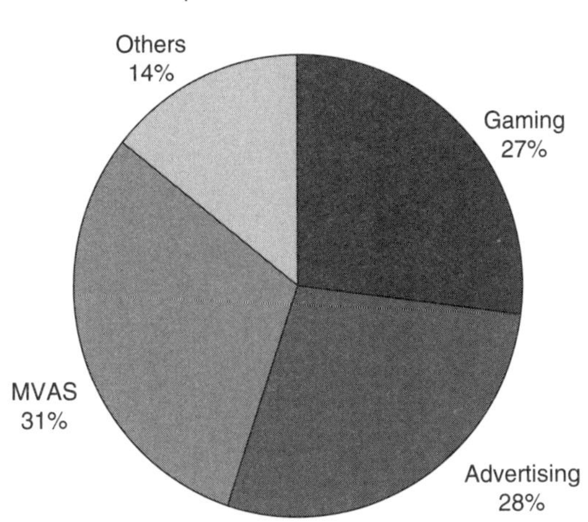

Source: Morgan Stanley research company data. Based on CQ3 05 earning results for Baidu, Shanda, NetEase, SINA, Tom online, Tencent, Solru, The9, Linktone, KongZhong, 51job, Ctrip, and Hurray, Companies selected based on market capitalization as of 2/13/06

Figure 14.14 China Internet Companies – Diversified Revenue Mix

to tap the potential of this new end market is addressing the wants and needs of Chinese consumers (to have more conveniences and to access information easily).

For instance, the proliferation of cell phones presents an opportunity (so far overlooked) for semiconductor companies capable of developing chips that enable value-added services such as e-mail and instant messaging, among others. Also, with the trend toward e-commerce likely to create rising demands for all types of online buying services, service providers will need processors that enable those applications.

The Chinese end market also has great opportunities for companies that can provide various types of enabling technologies for global markets. For example, the geographic size of the Chinese market creates demand for semiconductor solutions that enable efficient port connectivity across multiple networking and telecom sectors. Among the other technologies in demand are programmable silicon solutions: chips that integrate multiple protocols under software control; integrated line cards; chips enabling the convergence of networking and computing system design; solutions that accelerate connected computing; and solutions that can break access bottlenecks, among others.

Nowhere is the increasing importance of this market more evident than in the willingness of global investors to commit rising amounts of capital to non-US companies, in general, and to support Chinese technology innovation, in particular. In fact, many venture capitalist (VC) firms in the States have established funds specifically to take advantage of China's booming technology, gaming, broadband and cellular markets.

As the VC community invests in new US semiconductor ventures, it must be mindful of the need for these companies to be able to play in the Chinese market and the challenges that presents. New business models are emerging that must be addressed. For instance, having management teams located around the globe is becoming the rule, rather than the exception. When staffing new ventures, VCs must plan for and build leadership teams that are disparate from the outset. While it is an extra burden and challenge to have teams that are so widespread, the global approach has become critical to success, since it is imperative to have developers, managers and others located where the end markets are. In addition to providing funding and building global teams, VCs should also be prepared to assist their portfolio companies in identifying partners and potential acquisitions in the Chinese market.

But all of those who hope to play successfully in the new market will need more than just global teams and a physical presence in the country. They will also need a solid understanding of how the Chinese market differs from the western market culturally, intellectually, politically and otherwise.

China is deeply rooted in a different business and political history than the West. As the two markets merge, both sides must work together to develop an understanding regarding the importance of developing best business practices. The best practices that emerge in China may be different from the best practices in the United States. For example, Chinese companies are showing a determination to bootstrap their start-up businesses much, much longer than start-up companies in the United States. In fact, the fabless "miracle" that took place in Taiwan in the 1990s (in which small fabless companies succeeded by bootstrapping and then receiving early-stage funding as they expanded) is now being recreated in Beijing and Shanghai.

The ability to work effectively in a different culture is essential for US companies setting up shop in China. Many Chinese semiconductor start-ups are run by native Chinese engineers and entrepreneurs who have returned to the country after many years of studying, living and working in the United States. Their fluency with the languages and understanding of the business climates of both countries is invaluable in bridging East/West cultural differences and building market credibility.

It is important that US companies aiming for the Chinese end market take a page from this playbook. By building teams that include cross-cultural members who have an intimate knowledge of both Chinese and US business rules and values, these companies will be better able to successfully navigate any uncharted waters they may encounter.

Politically and historically, China may have roots in communism, but the success of its technology end market is proving its ability to foster and support capitalist businesses. The country has finally come into its own after years marked by very little industrial economic activity and growth (due to civil war and Japanese occupation from 1900 to 1950, a socialist experiment from 1950 to 1976, and privatization through the early 1990s). For the last 10 years, private companies have been developing organically in China. But although the country is experiencing a booming private sector focused on services and technology innovation, it is still in the very early stages of developing a coherent economic and social system (based on socialist state ownership of its core industries).

Funding differences between the West and China, while not a major issue, are nevertheless worth mentioning. The vast majority of funding activity in China is structured exactly as it is in the United States. Most VCs currently investing in China have been trained in the United States and are using the same vehicles (preferred stock) as US investors. But there are a few key funding differences. Investors and entrepreneurs should be aware that:

- To protect VC capital, most investments are made offshore in holding companies (which then own local People's Republic of China companies).
- It takes much less capital to build companies in China. For instance, because fabless companies in China require about 25 percent of the total capital required for similar US companies, smaller amounts of capital buy larger percentages of companies.
- There is little or no angel financing in China.

There is no doubt that the technology end market is undergoing a profound change. But while the base of power has moved, the opportunity remains. Currently, a number of American semiconductor companies are successfully selling into the new Chinese end market – and there is still plenty of room for new players to get into the game. For US-based semiconductor companies that take the global approach, understand and embrace the different business culture, and offer enabling technologies for the products and services in demand, a huge and willing market awaits.

Overall Semiconductor Revenues in 2006 (Public Companies)

Rank	Company (Fabless Bolded)	2006 (CY) Revenue (US$ 000)
1	Intel	$31,580,000
2	Samsung Electronics	$19,670,000
3	Texas Instruments	$13,730,000
4	Toshiba	$10,030,000
5	STMicroelectronics	$9,855,000
6	Renesas	$8,170,000
7	HYNIX	$7,375,000
8	Freescale	$6,080,000
9	Philips Semiconductor (NXP)	$5,935,000
10	NEC	$5,725,000
111	Micron Tech	$5,510,000
12	AMD	$5,245,000
13	Infineon	$5,055,000
14	Qimonda	$5,005,000
15	Sony	$4,742,000
16	Matsushita Electric Industrial	$4,450,000
17	IBM Microelectronics	$4,400,000
18	**QUALCOMM (QCT Division)**	**$4,331,000**
19	**Broadcom**	**$3,667,818**
20	Sharp	$3,490,000
21	Elpida	$3,450,000
22	**SanDisk Corporation**	**$3,257,525**
23	Fujitsu	$3,116,000
24	Hitachi	$3,089,100
25	**NVIDIA Corporation**	**$3,068,771**
26	ROHM	$2,945,000

(Continued)

Rank	Company (Fabless Bolded)	2006 (CY) Revenue (US$ 000)
27	Powerchip	$2,829,000
28	ADI	$2,595,000
29	Spansion	$2,580,000
30	Nanya Technology	$2,307,000
31	**Marvell Technology Group Ltd.**	**$2,237,596**
32	National Semi	$2,135,000
33	Maxim	$1,985,000
34	**LSI Logic**	**$1,982,148**
35	**Xilinx, Inc.**	**$1,871,604**
36	ProMos Technologies	$1,845,000
37	Atmel	$1,765,000
38	Fairchild Semiconductor	$1,655,000
39	Sanyo	$1,650,000
40	**MediaTek Inc.**	**$1,624,486**
41	Agere	$1,595,000
42	**Avago Technologies**	**$1,588,000**
43	Mitsubishi	$1,565,000
44	ON Semi	$1,525,000
45	Oki	$1,505,000
46	Vishay	$1,305,000
47	**Altera**	**$1,285,535**
48	International Rectifier	$1,171,120
49	Linear Technology	$1,140,000
50	Cypress	$1,092,000
51	Winbond	$1,059,000
52	Microchip Semi	$1,035,000
53	**Conexant Systems**	**$985,615**
54	**NovaTek**	**$964,314**
55	**Himax Technologies**	**$744,518**
56	Intersil	$741,000
57	**Cambridge Silicon Radio (CSR)**	**$704,700**
58	**VIA Technologies, Inc.**	**$657,901**
59	Macronix International	$587,000
60	**QLogic Corporation**	**$570,051**
61	**OmniVision Technologies, Inc.**	**$540,741**

(Continued)

Rank	Company (Fabless Bolded)	2006 (CY) Revenue (US$ 000)
62	Integrated Device Technology	$528,000
63	**Zoran Corporation**	**$495,846**
64	**Silicon Laboratories**	**$464,597**
65	**Silicon Storage Technology, Inc. (SST)**	**$452,092**
66	**PMC-Sierra, Inc.**	**$424,992**
67	**Phison**	**$382,080**
68	**Realtek Semiconductor Corporation**	**$381,191**
69	**Standard Microsystems Corporation (SMSC)**	**$365,499**
70	**MegaChips Corporation**	**$347,440**
71	**Power One**	**$338,048**
72	**Solomon Systech**	**$322,000**
73	**Etron Technology**	**$321,571**
74	**CoAsia Microelectronics**	**$310,187**
75	**Atheros Communications**	**$301,691**
76	**Silicon Image, Inc.**	**$294,958**
77	**Applied Micro Circuits Corporation (AMCC)**	**$289,678**
78	**Adaptec, Inc.**	**$276,688**
79	**Melexis NV**	**$265,841**
80	**Aeroflex (AMS Division)**	**$254,579**
81	**Semtech Corporation**	**$252,500**
82	**Wavecom**	**$249,028**
83	**SiRF**	**$247,680**
84	**Lattice Semiconductor Corporation**	**$245,459**
85	**Silicon Integrated Systems Corporation (SiS)**	**$242,682**
86	**Trident Microsystems, Inc.**	**$237,248**
87	**Genesis Microchip**	**$236,887**
88	**Integrated Silicon Solution, Inc. (ISSI)**	**$228,494**
89	**DSP Group, Inc.**	**$216,948**
90	**Wolfson Microelectronics Pls**	**$204,133**
91	**Core Logic**	**$202,452**
92	**Actel Corporation**	**$190,881**
93	**THine Electronics, Inc.**	**$181,429**
94	**Cirrus Logic, Inc.**	**$180,815**
95	**Elite Semiconductor Memory Technology**	**$178,490**
96	**Actions Semiconductor**	**$170,363**

(Continued)

Rank	Company (Fabless Bolded)	2006 (CY) Revenue (US$ 000)
97	**Faraday Technology**	**$170,114**
98	**Power Integrations, Inc.**	**$162,400**
99	**SigmaTel, Inc.**	**$159,365**
100	**AudioCodes**	**$147,353**
101	**Sitronix Technology Corporation**	**$137,803**
102	**ELAN Microelectronics Corporation**	**$136,821**
103	**Sirenza Microdevices, Inc.**	**$136,578**
104	**Ikanos Communications**	**$136,185**
105	**Pixelworks, Inc.**	**$133,607**
106	**Mindspeed Technologies, Inc.**	**$132,873**
107	**Richtek Technology**	**$132,096**
108	**Hittite Microwave**	**$130,290**
109	**Vimicro**	**$128,000**
110	**Gennum**	**$126,319**
111	**Mtekvision**	**$126,193**
112	**O2Micro International, Ltd.**	**$124,915**
113	**PixArt**	**$122,400**
114	**Holtek Semiconductor Ltd.**	**$119,392**
115	**Pericom Semiconductor**	**$118,803**
116	**Silan Microelectronics Joint-Stock Co., Ltd.**	**$113,000**
117	**Silicon Motion**	**$106,455**
118	**Monolithic Power Systems**	**$105,015**
119	**Leadis Technology**	**$101,208**
120	**Advanced Power Electronics**	**$101,013**
121	**ESS Technology, Inc.**	**$100,465**
122	**ALi Corporation**	**$100,061**
123	**NetLogic Microsystems**	**$96,806**
124	**Dialog Semiconductor**	**$94,024**
125	**Sigma Designs**	**$91,217**
126	**Sonix Technology**	**$86,100**
127	**Princeton Technology Corporation**	**$84,014**
128	**ZiLOG**	**$83,000**
129	**PLX Technology, Inc.**	**$81,425**
130	**Advanced Analogic Technologies**	**$81,161**
131	**Sipex**	**$78,750**

(Continued)

Rank	Company (Fabless Bolded)	2006 (CY) Revenue (US$ 000)
132	**DenMOS Technology (1995 revenue)**	$76,651
133	**Fidelix**	$75,801
134	**Volterra Semiconductor**	$74,588
135	**Tundra Semiconductor Corporation**	$71,923
136	**Exar Corporation**	$70,417
137	**California Micro Devices**	$69,936
138	**Anpec Electronics**	$69,469
139	**Microtune, Inc.**	$69,232
140	**Cheertek**	$66,861
141	**Emerging Memory & Logic Solution Inc.**	$66,727
142	**NIKO Semiconductor**	$66,616
143	**Telechips**	$65,993
144	**Centillium Communications, Inc.**	$64,563
145	**Catalyst Semiconductor**	$62,139
146	**Global Mixed-mode Technology Inc.**	$58,024
147	**Echelon Corporation**	$57,276
148	**Integrated Techology Express (ITE)**	$54,833
149	**Techwell, Inc.**	$53,712
150	**C&S Technology**	$52,688
151	**Myson Century Semiconductor, Inc.**	$51,611
152	**AMIC Technology Corporation**	$50,414
153	**Global Unichip Corporation (2005 revenue)**	$48,782
154	**Mellanox Technologies**	$48,539
155	**TLI**	$47,681
156	**Avid Electronics Corporation**	$47,008
157	**ENE Technology, Inc.**	$45,934
158	**Genesys Logic**	$45,812
159	**Nordic Semiconductor ASA**	$45,778
160	**CML Microcircuits Ltd. (2005 revenue)**	$42,598
161	**Hifn**	$42,361
162	**Ramtron International Corporation**	$40,481
163	**SmartASIC**	$40,380
164	**TranSwitch Corporation**	$38,920
165	**Weltrend Semiconductor, Inc.**	$38,294
166	**System General**	$38,018

(Continued)

Rank	Company (Fabless Bolded)	2006 (CY) Revenue (US$ 000)
167	FOCUS Enhancements	$37,478
168	Cosmo Electronics Corporation	$35,133
169	QuickLogic Corporation	$34,924
170	Irvine Sensors	$34,165
171	Pixelplus	$33,700
172	Ultrachip	$32,525
173	Alpha Microelectronics	$30,807
174	Brilliance Semiconductor	$30,746
175	Simtek	$30,630
176	Prolific Technology, Inc.	$29,641
177	C-Media Electronics Inc.	$28,015
178	Nanogen, Inc.	$26,852
179	Shanghai Fudan Microelectronics Co., Ltd.	$26,680
180	V-TAC Technology Co., Ltd.	$26,665
181	Analog Integrations	$25,407
182	RealVision, Inc.	$25,271
183	RDC	$24,241
184	Aimtron	$24,087
185	AME	$22,522
186	Silicon Touch	$21,234
187	Mosart Semiconductor	$21,141
188	Sanghwa Micro Tech.	$20,467
189	IndigoVision, Ltd.	$20,459
190	Jade Bird Universal Sci-tech Co.	$18,155
191	Makus	$17,462
192	Ours Technology	$15,710
193	Topshine Electronics	$15,434
194	Metalink	$14,476
195	Service & Quality	$13,225
196	Acard	$13,194
197	Davicom Semiconductor, Inc.	$12,795
198	Tripath Technology (2005 revenue)	$12,510
199	Tontek Design Technology Co., Ltd.	$10,034
200	TOPRO Technology Inc.	$9,972
201	Syntek Semiconductor	$9,727

(Continued)

Rank	Company (Fabless Bolded)	2006 (CY) Revenue (US$ 000)
202	**SwitchCore AB**	$8,665
203	**ASIX**	$8,131
204	**King Billion Electronics Co. Ltd.**	$7,671
205	**Tvia, Inc. (2005 revenue)**	$6,619
206	**Averlogic Technologies**	$5,891
207	**CombiMatrix Group**	$5,740
208	**MosChip Semiconductor**	$5,057
209	**LOGIC Devices, Inc.**	$4,901
210	**SATEC (formerly Aralion)**	$3,985
211	**ClearSpeed**	$3,811
212	**HiMARK Technology**	$3,375
213	**Innovision Research & Technology**	$3,262
214	**TM Technology**	$3,130
215	**Ever Electronics**	$2,731
216	**Apogee Technology Inc.**	$1,885
217	**TechnoConcepts**	$1,810
218	**ParkerVision (2005 revenue)**	$996
219	**NeoMagic Corporation** (Ending October 2006)	$559
220	**Cyan Holdings**	$527
221	**SpatiaLight, Inc.**	$474
222	**Rim Semiconductor** (Ending October 2006)	$62
223	**Mathstar**	$53
224	**Micromem Technologies Inc.** (Ending October 2006)	$10

Source: FSA.

Bibliography

Section 1.0 *A Decade of Transition in the Semiconductor Industry*, *The Fabless Business Model, Fabless Forum*, December 2004, Mark Edelstone, Managing Director, Morgan Stanley.

Section 1.6.1 *Semiconductor Firms are Forging a New Path*, *Fabless Forum*, July 1998, Jean-Luc Pelissier, Vice President and General Manager, SABER.

Section 1.7 *A Case Study: Cypress Semiconductor*, *Fabless Forum*, December 1998, Christopher Norris, VP of Programmable Logic, Cypress Semiconductor Corporation.

Section 1.9 *Geographic Manufacturing Centers,* CIBC World Markets, Industry Research, Semiconductors, December 2005.

Section 3.0 *Qualities of a Successful Fabless Company*, Fabless Forum, December 2002, Steven Eliscu, Alliance Semiconductor.

Section 4.0 *Fabless ASIC Model,* Contribution for *Understanding Fabless IC Technology,* eSilicon Corporation.

Section 5.2 *Semiconductor Manufacturing Steps,* www.sematech.org, Sematech, Sematech.

Section 6.0 *Electronic Design Automation*, Contribution for *Understanding Fabless IC Technology*, Synopsys.

Section 6.0 *Electronic Design Automation*, Contribution for *Understanding Fabless IC Technology*, Mentor Graphics.

Section 6.0 *Electronic Design Automation*, Contribution for *Understanding Fabless IC Technology*, Bruce Chan, Pankaj Mayor, Cadence.

Section 6.1 *Fabless EDA Overview*, Contribution for *Understanding Fabless IC Technology*, Cadence.

Section 6.3 *Physical Design EDA*, *Fabless Forum*, September 2001, David Gregory, CEO, ReShape, Inc.

Section 7.21 *The Evolution of the IP industry, Fabless Forum,* March 1999, Jeff Lewis, Vice President of Marketing, Artisan Components, Inc.

Section 7.22 *Intellectual Property Consideration,* Contribution for *Understanding Fabless IC Technology,* ARM.

Section 7.23 *IP Outsourcing, Fabless Forum,* September 2001, Adam Kablanian, CEO, Virage Logic.

Section 7.24 *Making IP Work in the Fabless Semiconductor Community,* Contribution for *Understanding Fabless IC Technology,* Jim Ensell, Vice President of Business Development and CIO, eSilicon Corporation.

Section 7.25 *IP Acquisition Considerations for Fabless IC Companies, Fabless Forum,* December 2003, Peter Lee, President & CEO, Aplus Flash Technology.

Section 8.1 *The Virtual Fab Challenge*, Section 8.3 *"Build to Forecast" for Outsourced Manufacturing*, Section 8.5 *The Information Ecosystem, Where Communication is Key*, Indu Navar, Founder & CEO, Serus Corporation.

Section 8.2 *Semiconductor & Fabless Manufacturing – What is Different?* Ron Jones, Founder & CEO, N-able Group.

Section 9.0 *Quality and Reliability*, Contribution for *Understanding Fabless IC Technology*, Evert A. Wolsheimer Dr., Co-author.

Section 10.1 *Simplifying Outsourced Test Development*, *FSA Forum*, September 2005, Allan Calamoneri, VP of Test Business Development, Carsem.

Section 11.1 *Achieving Best-in-Class Operations Practices*, *FSA Forum*, December 2005, Rakesh Kumar Dr., President, TCX Inc.

Section 11.2 *A Foundry Manager's Role in a Fabless Company*, *Fabless Forum*, December 1998, Rajan Saigusa, Sr. Foundry Engineer Manager, SEEQ Technology.

Section 11.3 *Closing the Loop: Understanding the Manufacturing Flow*, *Fabless Forum*, March 2003, William Miller, Vice President, Chief Architect, SiVerion, Inc.

Section 11.4 *Managing a Virtual Manufacturing Chain*, *Fabless Forum*, July 1998, Kurt Milne, Director of Industry Marketing, Camstar Systems.

Section 12.2 *Best Practices: Operations in a Fabless Start-Up,* Contribution for *Understanding Fabless IC Technology,* Gina Gloski, President, Semiconductor Operations Consulting.

Section 12.3 *Legal Issues for Fabless Semiconductor Companies*, George McKinnis, Legal Counsel, TranSwitch Corporation.

Section 12.4 *Semiconductor Back-End Subcontracting No Longer a Zero-Sum Game*, *Fabless Forum*, March 2001, Jim Healy, President, ASAT, Inc.

Section 13.1 *Creating Successful Corporate Boards in Fabless Companies*, *Fabless Forum*, December 2002, Tim O'Shea, Partner-in-Charge, Heidrick & Struggles, Inc.

Section 13.2 *Finding the Right VC, Fabless Forum*, December 2002, Kenn W. Webb, Chairman, Corporate-Securities Section, Davis Munck, P.C.

Section 14.1 *Keeping Up with the Pace of Change in a Fabless World, FSA Forum*, March 1999, Rob Hilkes, Director of Marketing, MOSAID Semiconductor.

Section 14.2 *Foundry Roadmaps: Partnering, Leading and Innovating, FSA Forum*, September 2005, Paul Kempf, Chief Technology & Strategy Officer, Jazz Semiconductor.

Section 14.3 *Semiconductor Manufacturing in the 21st Century*, Contribution for *Understanding Fabless IC Technology*, Ajit Manocha, Executive VP and Chief Manufacturing Officer, NXP, Eindhoven, The Netherlands.

Section 14.4 *The Emerging Dominance of China in Technology and End Markets, FSA Forum*, September 2006, Dave Furneaux, Founder & Managing General Partner, Kodiak Venture Partners.

Other Credits:

Figure 1.6, UMC, Wafer Fab Photos

Figure 2.5, ASE, Flip Chip Technology Photos

Figure 2.6, ASE, Wafer Bumping Photos

Figure 2.7, ASE, 300mm Wafer Backend Assembly & Testing Photos

Collaboration / Common Platform, Chartered Semiconductor Manufacturing, Contribution for Understanding Fabless IC Technology.

Shipping, Contribution for Understanding Fabless IC Technology, Scott Berlin, JSI Shipping.

Specialty Fabs, Contribution for Understanding Fabless IC Technology, Volker Herbig, Thomas Hartung, X-Fab.

Glossary of Terms and Acronyms

Analog Circuit:	Electronic circuit that operates with currents and voltages that change continuously without abrupt transitions between levels.
Analog IC:	A semiconductor device used for electrical signal processing, power control or electrical drive capability in which some of the inputs or outputs can be defined in terms of continuously or linearly variable voltages, currents or frequencies.
ADC:	Analog-to-Digital Converter
Application-Specific Integrated Circuit (ASIC):	A chip that is custom designed for a specific application and does not have to retrieve and interpret stored instructions.
Application-Specific Standard Product (ASSP):	An integrated circuit that is dedicated to a specific application and sold to more than one user. It is manufactured using vendor-supplied tools and libraries for use in a distinct electronic equipment category.
ASIC:	*See* Application-Specific Integrated Circuit.
Assembly:	A procedure in which a fully processed semiconductor device is mechanically and electrically connected to the package.
ASSP:	*See* Application-Specific Standard Product.
ATE	Automatic Test Equipment
Back-End Manufacturing:	The testing and assembly of chip manufacturing that occurs after the wafer has left the clean room.
Back-end:	Packaging and testing procedures during chip fabrication. Also, back-end refers to the IC physical design process, including layout and packaging design.
BiCMOS:	*See* Bipolar Complementary Metal Oxide Semiconductors.
Bill-of-Material (BOM):	A document that describes a product and includes a list of the parts, tools, drawing notes and materials required for the manufacturing of that product.
Bipolar Complementary Metal Oxide Semiconductor (BiCMOS):	Bipolar complementary metal oxide semiconductor (also referred to as BiMOS) refers to the integration of bipolar junction transistors and CMOS technology into a single device.

BISR: *See* Built-In Self-Repair.
BIST: *See* Built-In Self-Test.
BOM: *See* Bill-of-Material.
Built-In Self Test (BIST): A test procedure using the inherent components of a device or system to evaluate its ability to perform an expected function.
Built-In Self-Repair (BISR): An integrated circuit semiconductor device with a built-in circuit for embedded memory capable of repairing its memory.
Capacity Utilization: The relationship between the actual output produced and the potential output that could be produced with installed equipment.
Capital Investment: Money paid to purchase a capital asset or a fixed asset.
Chemical Mechanical Planarization (CMP): The process of using chemical reactions to polish the surfaces of wafers and films during semiconductor fabrication.
Chemical Vapor Deposition (CVD): A chemical process used to produce high-purity and high-performance solid materials.
Chipset: A number of integrated circuits designed to perform one or more related functions.
Circuit Design: Design of circuit logic that allows electronic components to perform a specific function.
CMOS: *See* Complementary Metal Oxide Semiconductor.
CMP: *See* Chemical Mechanical Planarization.
Complementary Metal Oxide Semiconductor (CMOS): A type of semiconductor that uses negative and positive polarity circuits and requires less power than a chip using only one type of transistor, since only one of the circuits is on at the same time.
Complex Programmable Logic Device (CPLD): A programmable integrated circuit configured for digital hardware, such as mobile phones, that can handle larger designs than simple programmable logic devices, but less logic than field programmable gate arrays.
Conductor: A material that efficiently transfers and electrical charge because it has an excess of unbound electrons and easily gives them up.
Core Memory: The main memory in a computer.
Cost Target: The maximum allowable expenditure for material, labor, outsourcing, overhead and all other expenses associated with project.
Cost: The total money, time and resources associated with a purchase or activity.
COT: *See* Customer Owned Tooling.
Coupling: To electronically connect two circuits so that signal will pass from one to the other.

CPLD: *See* Complex Programmable Logic Device.

CRM: *See* Customer Relationship Management.

Customer Owned Tooling (COT): Customers provide their chip design to a foundry for manufacturing and because they also pay an NRE fee for the manufacturing masks and reticles, it is said that the customer owns the tooling.

Customer Relationship Management (CRM): Methodologies and software that help an enterprise manage customer relationships in an organized manner.

CVD: *See* Chemical Vapor Deposition.

Cycle Time: The total time to move a unit of work from the beginning to the end of a process.

DAC: Digital-to-Analog Converter

Demand: The amount of a particular economic good or service that consumers are willing to purchase at a given price.

Design for Manufacturing (DFM): The general engineering art of designing products in such a way that they are easy to manufacture. DFM includes a set of techniques to modify the design & integrated circuits to make them more manufacturable.

Design for Test (DFT): A design technique that adds test features to a microelectronic hardware product design to validate that the hardware contains no defects that might adversely affect product functioning.

Design IP: Design of circuit logic that allows electronic components to perform a specific function.

Design Services: Semiconductor technology and design engineering resources to create a variety of advanced integrated circuits.

DFM: *See* Design for Manufacturing.

DFT: *See* Design for Test.

Dicing: Mechanical process of cutting a semiconductor wafer into squares or rectangles during manufacturing.

Die (Chip): A square or rectangular piece cut from a wafer that contains an electrical pattern repeated in several rows and columns across the top surface of the wafer.

Dielectric: An insulator such as glass, rubber or plastic that is able to retain an electrostatic charge without current flowing through them.

Digital Signal Processing (DSP): The manipulation of analog information, such as sound or photographs, which has been converted into a digital form using a data compression technique.

Digital Subscriber Line (DSL): A high-speed digital switch service using existing copper pairs to connect subscribers to a central office that handles more data flowing toward the subscriber than toward the network.

Digital: Circuit design based on the binary number system.

Discrete: A semiconductor or semiconductor die that contains only one active device, such as a transistor or a diode.

DRAM: *See* Dynamic Random Access Memory.

DSL: *See* Digital Subscriber Line.

DSP: *See* Digital Signal Processor.

Dynamic random access memory: Memory access that allows the computer's processor to access any part of the memory directly rather than having to proceed sequentially from a starting place. Its storage cells must be refreshed with a new electronic charge every few milliseconds.

Economic Value Added (EVA): The monetary value of an entity at the end of a time period minus the monetary value of that same entity at the beginning of the same time period.

EDA: *See* Electronic Design Automation.

EEPROM Memory: *See* Electrically Erasable Programmable Read-Only.

EHS: *See* Environmental, Health, and Safety.

Electrically Erasable Programmable Read-Only Memory (EEPROM): Electrically erasable programmable read-only memory can be erased by exposing it to an electrical charge and retains its contents after the power has been turned off.

Electronic Design Automation (EDA): The category of tools for designing and producing electronic systems ranging from printed circuit boards to integrated circuits which is sometimes referred to as electronic computer-aided design or CAD.

Electronics Manufacturing Services (EMS): A term used to refer to companies that design, test, manufacture, distribute and provide return and repair services for electronic component and assemblies for original equipment manufacturers.

EMS: *See* Electronics Manufacturing Services.

Enterprise Resource Planning (ERP): A management information process that a company manages and integrates the important components of its business, such as planning, purchasing, inventory, sales, marketing, finance and human resources.

Environmental, Health, and Safety (EHS): The category of procedures and regulations related to maintaining environmental quality standards for health and safety, which includes the Restriction of Hazardous Substances (RoHS) and Waste Electrical and Electronic standards (WEES).

Epitaxial Wafer: Consists of a thin, single-crystal layer grown on the polished surface of the basic wafer substrate. The substrate, which is designed to have different composition and electrical properties from the layer of single-crystal silicon on the wafer surface, among other things, helps to improve isolation between circuit elements fabricated on the silicon film surface of the wafer.

ERP: *See* Enterprise Resource Planning.

Etch: A lithographic process that removes material from selected areas of a die.

EVA: *See* Economic Value Added.

Fab: Short for fabrication facility. *See* Fabrication Facility.

Fabless: A company that does not manufacture its own silicon wafers and concentrates on the design and development of semiconductor chips.

Fab-Lite (Hybrid Strategy): A company that owns and operates its own semiconductor fabrication facilities, but also farms out some of its manufacturing to outside foundries.

Fabrication Facility: A manufacturing plant that makes semiconductor devices. A fab contains at least one processing line to do wafer patterning, doping, deposition and etching to manufacture semiconductor devices.

Failure Analysis: Logical and systematic examination of an electronic circuit, machine, or equipment, and its documentation to detect and analyze the causes, probabilities, and consequences of failure.

Failure Modes and Effects Analysis (FMEA): Methodology for analyzing potential reliability problems early in the development cycle when it is easier to take action to prevent problems that could result in failure of a product.

Field Programmable Logic Device (FPLD): A field programmable Logic Device is distributed without programming by the manufacturer, so that it may be programmed by a user to execute a desired function.

Field Programmable Gate Array (FPGA): A semiconductor device containing programmable logic components called "logic blocks", and programmable interconnects. Logic blocks can be programmed to perform the function of basic logic gates such as AND, and XOR, or more complex combinational functions such as decoders or simple mathematical functions. In most FPGAs, the logic blocks also include memory elements.

Flip Chip: An electronic component or semiconductor device that can be mounted directly onto a substrate, board, or carrier facedown.

FMEA: *See* Failure Modes and Effects Analysis.

Foundry: A semiconductor manufacturer that makes chips for third parties.

FPGA: *See* Field-Programmable Gate Array.

FPLD:	*See* Field Programmable Logic Device.
Front-End Processing:	Manufacturing process for the formation of transistors directly on silicon.
GaAs:	*See* Gallium Arsenide.
Gallium Arsenide (GaAs):	An alloy of gallium and arsenic compound that is used as the base material for chips.
Gate Array:	A type of application-specific integrated chip that is partially finished with rows of the transistors and resistors built in but unconnected. The chip is completed by designing and attaching the top metal layers that provide the interconnecting pathways.
GDSII:	*See* Gerber Data Stream Information Interchange.
General-Purpose Logic Chip:	A general-purpose logic chip (also called a microprocessor) processes data by following instructions in a software program. Microprocessors are flexible chips, since software may be changed.
Gerber Data Stream Information Interchange (GDSII):	A database format that has been the standard for integrated circuit layout data exchange developed by Calma for the company's layout design computer systems Graphic Data system (GDS) and GDSII and now owned by Cadence Design Systems.
Hardware:	Refers to objects that may be touched, such as disks, disk drives, display screens, keyboards, printers, boards, and chips.
Hybrid Strategy:	*See* Fab-Lite.
IDM:	*See* Integrated Device Manufacturer.
Indium Phosphide:	A binary semiconductor composed of indium and phosphorus. It is used in high-power and high-frequency electronics because of its superior electron velocity with respect to the more common semiconductors silicon and gallium arsenide.
Ingot:	A cylindrical piece of semiconductor material from which individual wafers will be sliced.
InP:	*See* Indium Phosphide.
Integrated Device Manufacturer (IDM):	A company that performs every step of the chip making process, including design, manufacture, test, and packaging. Examples of IDMs are Intel, AMD, Freescale, IBM, TI, and Lucent.
Intellectual Property (IP):	An intangible asset that consists of human knowledge and ideas, such as patents, copyrights, and trademarks.
Interconnect:	To attach one device to another.
International Organization for Standardization	Organization acting as a central clearinghouse for industry standards drafted by national standard setting organizations.

ISO:	The American National Standards Institute is the U.S. Representative of ISO.
	A method for chrome plating and other plating procedures, as well as a process that may improve engineering properties of substrate materials.
Ion Implantation:	A materials engineering process by which ions of a material can be implanted into another solid, thereby changing the physical properties of the solid.
	An atom or group of atoms which have lost or gained one or more electrons, making them positively or negatively charged.
IP:	*See* Intellectual Property
ISO 9000 Certification:	ISO certification ensures that the processes to develop a product are documented and performed in a quality manner. The certification process involves a plan for documenting the company's ISO system followed by interviews with the company's management and line staff to make sure that the new system has been effectively implemented.
ISO 9001:	ISO 9001 provides standards and guidelines of the International Organization for Standardization (ISO) related to the quality in design and development by manufacturing and service industries.
ISO:	*See* International Organization for Standardization.
Joint Venture (JV):	A contractual agreement joining two or more parties for the purpose of executing a particular business undertaking where all parties agree to share in the profits and losses.
JV:	*See* Joint Venture.
Leakage:	Small undesirable flow of current through an insulator or dielectric.
Lithography:	The process of imprinting patterns on semiconductor materials to be used as integrated circuits.
Mask Set:	A mask set is a series of electronic data that defines geometry for the photolithography steps of semiconductor fabrication.
Materials Requirement Planning (MRP):	Computer-based information system designed to handle ordering and scheduling of inventory, such as raw materials, component parts, and subassemblies that will be used in the production of a finished product.
Memory:	Data storage in the form of chips.
MEMS:	*See* Microelectromechanical Systems.
Microcontroller:	A single chip that contains the processor (CPU), nonvolatile memory for the program (ROM or flash), volatile memory for

input and output (RAM), a clock and an I/O control unit. Also called a computer on a chip.

Microelectromechanical Systems (MEMS): The integration of mechanical elements, sensors, actuators, and electronics on a common silicon substrate.

Micrometer: A micrometer (μm) is a unit of length equal to one millionth of a meter and is also commonly known as a micron.

Micron: Another term for micrometer. *See* Micrometer.

Microprocessor: A microprocessor is a silicon chip that contains a CPU. The terms microprocessor and CPU are used interchangeably.

Mixed-Signal Integrated Circuit: A mixed-signal integrated circuit combines analog and digital circuitry on a single semiconductor die.

Moore's Law: The observation made in 1965 by Gordon Moore, co-founder of Intel, that the number of transistors per square inch on integrated circuits doubled each year since the integrated circuit was invented.

MRP: *See* Materials Requirement Planning.

NAND Flash Memory: The preferred format for storing large quantities of data on devices such as USB Flash drives, digital cameras and MP3 players. Higher density, lower cost, and faster write and erase times, and a longer rewrite life expectancy make NAND especially well suited for consumer media applications where large files of sequential data need to be loaded into memory quickly and replaced with new files repeatedly.

Nanometer: A Nanometer is one billionth of a meter.

Netlist: A computer file or a printed listing that contains a list of the signals in an electronic design with all of the circuit elements (transistors, resistors, capacitors, and ICs.) connected to that signal in the design.

Noise: Noise refers to random, unpredictable, and undesirable signals, or changes in signals, that mask the desired information content.

Non-Disclosure Agreement (NDA): A non-disclosure agreement is a signed formal agreement in which one party agrees to give a second party confidential information about its business or products and the second party agrees not to share this information with anyone else for a specified period of time.

Non-Recurring Expense (NRE): A cost of doing business resulting from revenue-generating activities occurring only once on a company's financial statement.

NOR Flash Memory: Memory used to store small amounts of executable code for embedded computing devices, such as PDAs and cell phones, and is well suited for code storage, because of its reliability, fast read operations, and random access capabilities. Since code can be directly executed in place, NOR is used for storing firmware, boot code, operating systems, and other data that rarely changes.

NRE: *See* Non-Recurring Expense.

ODM: *See* Original Device Manufacturer

OEM: *See* Original Equipment Manufacturer.

OM: *See* Operations Management.

Operations Management (OM): An area of business that is concerned with the production of services and products, including the management of resources and distribution to customers.

Original Device Manufacturer (ODM): An equipment vendor to an end user that usually designs the equipment, and takes on some aspects of the chip/system design process.

Original Equipment Manufacturer (OEM): An equipment vendor to an end-user that usually designs the equipment but is not the manufacturer. A software company that sells products incorporated into PCs may refer to its customer, the PC hardware vendor, as an OEM.

OSAT: *See* Outsourced Assembly and Test.

Outsourced Assembly and Test (OSAT): An arrangement where one company performs services for another which includes assembly and testing procedures to verify that the wafers have not been damaged during processing procedures.

Oxidation: Any chemical reaction in which a material gives up electrons when the material combines with oxygen. Burning is an example of rapid oxidation; rusting is an example of slow oxidation.

PA: *See* Power Amplifier.

Packaging: The protective container or housing for an electronic component or die, with external terminals to provide electrical access to the components inside. Packages provide power and signal distribution, power dissipation, and physical and chemical protection of the circuits.

PCB: *See* Printed Circuit Board.

PDK: *See* Process Design Kit.

PDM: *See* Product Data Management.

Photolithography: A photographic process that is used to transfer circuit patterns onto a semiconductor wafer in the manufacture of computer chips. Beams of light are projected through a patterned reticle onto a silicon wafer that is covered with a photosensitive material that etchs a circuit into the semiconductor wafer.

Photomask: A film or glass negative that has many high-resolution images used in the production of semiconductor devices and integrated circuits.

Photoresist: A radiation-sensitive material that masks portions of the substrate.

PLD: *See* Programmable Logic Device.

Power Amplifier (PA): The final stage in multistage amplifiers, such as audio amplifiers and radio transmitters, designed to deliver maximum power to the load with a given percent of distortion.

Printed Circuit Board (PCB): A rigid, flat board that holds chips and other electronic components. The board is made of layers, typically 2–10 that interconnect components via copper pathways. The main printed circuit board in a system is called the system board or motherboard, while the smaller ones that plug into the slots in the main board are called boards or cards.

Probe Station: Used for making contact to microscopic features on a semiconductor device.

Process Design Kit (PDK): A set of data files that enable analog circuit and layout designers to design an integrated circuit using a set of electronic design automation tools and a selected foundry process.

Process Technology: Method used to make silicon chips.

Product Change Notification (PCN): Document that informs the customer of a process change.

Product Data Management (PDM): The management and classification of design data and specifications for an engineered product and the management of change to this information.

Product Data Management (PDM): An information system used to manage the data for a product as it passes from engineering to manufacturing, including plans, drawings, images, and notes.

Programmable Logic Device (PLD): An integrated circuit that can be programmed in a laboratory to perform complex functions that includes arrays of AND and OR gates.

R&D: *See* Research and Development.

Radio Frequency (RF): Any frequency within the electromagnetic spectrum associated with the travel of radio waves. When a radio frequency current is supplied to an antenna, an electromagnetic field is created that will be able to travel through space.

Register Transfer Level (RTL): A hardware description language for defining digital circuits. The circuits are described as a collection of registers, Boolean equations, control logic such as if-then-else statements as well as event sequences. Popular RTL languages are VHDL and Verilog.

Research and Development (R&D): The process of discovering new knowledge about products, processes, and services, and then applying that knowledge to create new and improved products, processes, and services that fill market needs.

Reticle: A grid or pattern placed in the eyepiece of an optical instrument, used to establish scale or position.

Return Material Authorization (RMA): Transaction where the recipient of a product arranges to return defective goods to the supplier to have the product repaired or replaced or in order to receive a refund or credit for another product from the same corporation.

Return on Invested Capital (ROIC): A measure of how effectively a company uses money invested in its operations.

RF: *See* Radio Frequency.

RMA: *See* Return Material Authorization.

ROIC: *See* Return on Invested Capital.

RosettaNet: An independent, self-funded, nonprofit group dedicated to the development and deployment of standard electronic business interfaces.

RTL: *See* Register Transfer Level.

Sales Force Automation (SFA): A technique of using software to automate the business tasks of sales, including order processing, contact management, information sharing, inventory monitoring and control, order tracking, customer management, sales forecast analysis, and employee performance evaluation.

Sarbanes-Oxley: The Sarbanes-Oxley Act of 2002 is legislation enacted in response to the Enron and WorldCom financial scandals. The act is administered by the Securities and Exchange Commission, and defines which records are to be stored and for how long. The consequences for non-compliance are fines, imprisonment, or both.

SCM: *See* Supply Chain Management.

Semiconductor: A semiconductor is an element that has an electrical conductivity in a range between conductors and insulators. Integrated circuits are typically fabricated in semiconductor materials such as silicon, germanium, or gallium arsenide.

Semiconductor Manufacturer: A semiconductor manufacturer designs, develops, manufactures, and markets integrated circuits.

Semiconductor Intellectual Property (SIP): An intangible asset that consists of human knowledge and ideas, such as patents, copyrights, and trademarks specifically used in the manufacture of semiconductors.

SFA: *See* Sales Force Automation.

Shuttle Services: The consolidation of a number of customer designs on a single mask, which should produce wafers hosting multiple customers' chips in quantities sufficient to allow customer sampling. A shuttle schedule manages launch times to keep track of varying designs coming in from multiple customers, and a design must be submitted by its launch date to retain its reserved shuttle seat.

SiGe: *See* Silicon Germanium.

Silicon Germanium (SiGe): A semiconductor material made from silicon and germanium. Germanium is very similar to silicon, but when one layer is grown on top of the other to form the base of the transistor, the resulting transistor can switch faster and yield higher performance.

Silicon: Silicon is a chemical element that is brittle with a metallic luster. It is the most abundant electropositive element in the Earth's crust.

SiP: *See* System in Package.

SIP: *See* Semiconductor Intellectual Property

SOC: *See* System-on-a-Chip.

Software: Instructions or data that can be stored electronically.

SPC: *See* Statistical Process Control.

Sputter Deposition: Method of depositing thin films by sputtering a block of source material onto a substrate. Sputtering is used in the semiconductor industry during integrated circuit processing.

SRAM: *See* Static Random Access Memory.

Standard Cell: Predefined circuit elements that may be selected and arranged to create a custom or semi-custom integrated circuit more easily than through design. Designers build ASICs using standard cells.

Static Random Access Memory (SRAM): Static random access memory is random access memory (RAM) that retains data bits in its memory as long as power is being supplied. SRAM does not have to be periodically refreshed. Static RAM provides faster access to data.

Statistical Process Control (SPC): The use of statistical methods to analyze a process or output to take appropriate actions to achieve and maintain a state of statistical control and continuously improve the process capability.

Substrate: The base or supporting materials to which layers or materials are applied. It is the part of the wafer from which dies are cut and is the electrical grounding for circuits. Semiconductor substrates are usually made from silicon.

Supply Chain Management (SCM):	Procedure for maintaining control of materials, information, and finances as they move in a process from supplier to manufacturer to wholesaler to retailer to consumer.
Synthesis (Logic Synthesis):	A computer process that transforms a circuit description from one level of abstraction to a lower level, usually towards some physical implementation. Logic synthesis was previously called hardware compilation.
System in Package (SiP):	A complete system packaged in one housing. A SiP contains several integrated circuits including a microprocessor on a single substrate such as ceramic or laminate. A SiP is really a multi-chip module that contains all the parts of a complete system.
System-on-a-Chip (SOC):	The electronics for a complete, working product contained on a single chip that includes the computer and all required electronics.
Tape-out:	The final stage of the design of an integrated circuit. It is the point at which the description of a circuit is sent to the manufacturer.
Technology Roadmap:	A technique for a product and technology planning process. Technology companies rely on roadmaps to connect product and technology plans and map them to market opportunities.
Thermal Diffusion:	A phenomenon in which a temperature gradient in a mixture of fluids gives rise to a flow of one constituent relative to the mixture as a whole.
Transistor:	A small electronic device containing a semiconductor that has at least three electrical contacts used in a circuit, such as an amplifier, detector, or switch.
Turnkey:	A complete system of hardware and software delivered to the customer ready to operate.
VC:	*See* Venture Capital.
Venture Capital (VC):	Funds made available for startup firms and small businesses with exceptional growth potential.
Verification:	The task of establishing the correctness of a design using electronic design automation tools to automatically check the timing, connections, and rules used to design the circuit.
Verilog:	Verilog is a hardware description language developed by Gateway Design Automation (now part of Cadence) in the 1980s.
VHDL:	VHDL (VHSIC Hardware Description Language) is a hardware description language developed in the 1980s by IBM, Texas Instruments, and Intermetrics under US government contract

	for the Department of Defense's VHSIC (Very High Speed Integrated Circuit) program.
Wafer Bumping:	The process of forming ball interconnections on a wafer prior to dicing. Two commonly used wafer bumping methods are screen deposition and electroplating.
Wafer Level Reliability (WLR):	A methodology to assess the reliability impact of tools and processes by testing mechanism-specific test structures under accelerated conditions during device processing.
Wafer Processing:	Processing consists of depositing material uniformly across the wafer at a controlled thickness (deposition), coating the wafer with photo-resist and pattern it using light (or x-ray or electron beam) and develop it to leave a negative or positive image of the desired pattern (patterning), and using chemicals (such as acids) to remove the material that is not needed (etching).
Wafer Sort:	A process where the integrated circuitry on each specific die is electrically tested with computer-controlled probes. Each wafer may contain up to hundreds of separate dies or chips that are tested. Devices that fail the test are marked with a colored dye and sorted accordingly.
Wafer Testing:	A process performed during semiconductor device fabrication where equipment is used to verify that the wafers have not been damaged during previous processing procedures.
Wafer:	A silicon wafer is a thin, circular slice of single-crystal semiconductor material used in manufacturing of semiconductor devices and integrated circuits.
Wireless LAN (WLAN):	A local area network that transmits over the air typically in the 2.4 GHz or 5 GHz unlicensed frequency band and does not require line of sight between sender and receiver. Wireless base stations are wired to an Ethernet network and transmit a radio frequency over an area of several hundred feet through walls and other nonmetal barriers. Roaming users can be handed off from one access point to another like a cellular phone system.
Yield Management:	In the semiconductor industry, it is the process used to maximize the number of effective finished goods (EFGs) by minimizing defects throughout the entire fabrication, packaging, and test processes.
Yield:	The percentage of chips in a finished wafer that pass all tests and function properly.

About the Authors

Jorge ("Jeorge") S. Hurtarte, P.E. and Ph.D Fellow is an executive with over 21 years of semiconductor industry experience in design engineering, manufacturing operations and general management.

Hurtarte is currently senior vice president at TranSwitch Corporation (Shelton, Connecticut), a fabless semiconductor company focused on the design and marketing of solutions for the telecommunications and data communications markets. His responsibilities at TranSwitch have included all worldwide VLSI chip design, software and technology development, EDA and manufacturing supplier selection, supply-chain management and manufacturing logistics, and corporate strategic programs.

Prior to TranSwitch, Hurtarte spent 15 years at Rockwell Semiconductor Systems, now Conexant Systems. During his tenure there, he progressed from the position of semiconductor product engineer to director of an engineering group responsible for developing ASSPs, module products and reference designs for data communications semiconductor products. Most recently at Rockwell, Hurtarte served as vice president, responsible for a business P&L and management of all related engineering and manufacturing operations resources, including a manufacturing plant.

Hurtarte earned a bachelor's degree in electrical engineering from the University of California, Irvine, and a master's degree in business administration from Auburn University, and is also a Ph.D Fellow at Auburn University. Hurtarte is also a graduate from Harvard Business School's Advanced Management Program and a Registered Professional Engineer in the State of California.

Hurtarte served on the Board of Directors of FSA from 2002 to 2005 and he is also on the Advisory Board of TUV Rheinland of North America. At FSA, Hurtarte has served for 3 years in the Technology, Education and Finance committees.

Dr. Evert A. Wolsheimer has been in the semiconductor industry for over 20 years. Evert worked as Vice President of Quality and Reliability at Xilinx, one of the largest and most successful fabless companies. At Xilinx, Wolsheimer previously served as Vice president and general manager of Xilinx's CPLD Division, overseeing product planning, design, software, operations and marketing. Before that he was Vice President of product technology, with responsibility for product engineering, process and packaging technology development, semiconductor foundry relationships and reliability. He joined the company in 1991 as director of technology development.

Prior to Xilinx, Wolsheimer was manager of technology development at LSI Logic and held several marketing and R&D management positions at Philips. He has three US patents at Xilinx and has published numerous papers for industry conferences and technical trade journals. Wolsheimer served on the Board of Directors of FSA from 1996 to 2005.

Wolsheimer earned a doctorate in electrical engineering from Delft University in the Netherlands.

Lisa Tafoya brings more than 16 years of experience in market research, consulting, project management and marketing communications to her role with FSA. Her in-depth ability to assess industry trends, create meaningful analysis and creatively promote various research activities make her a valuable asset as FSA's Vice President of Global Research.

In her role, Ms. Tafoya performs the strategic planning and oversees the implementation of all FSA member reports, surveys and deliverables. She oversees all data collection and analysis, as well as global publications created and distributed by the Association, including its quarterly Fabless Fundings and Global Fabless Financial Reports and the Outsourcing and Supplier Directories. She also manages all survey activities, such as quarterly Wafer Pricing, Back-End Pricing and Design Innovation Benchmarking surveys, as well as executive surveys distributed industry-wide to examine industry and business trends.

Ms. Tafoya also serves as the executive editor of FSA's industry-leading journal, *FSA Forum*. A quarterly magazine that examines hot topics in the semiconductor industry, *FSA Forum* is the premier magazine exclusively targeting the fabless, IDM and outsourcing markets and its supply-chain partners.

In addition, Ms. Tafoya is responsible for all of FSA's numerous committees worldwide. She manages their strategic focus and oversees the committee deliverables, including FSA's IPecosystem Tool Suite, which includes the Hard IP Quality Risk Assessment Tool; MS/RF Spice Model Checklists and PDK Checklists; the Standard Foundry Process Qualification Guideline (now a standard under JEDEC); as well as industry events; reports and other standards activity.

Tafoya earned a bachelor's degree in marketing from Southern Methodist University in Dallas, Texas.

About FSA

FSA is the voice of the global fabless business model. Industry leaders incorporated FSA in 1994 on the premise that the fabless business model would be a viable, long-term business model. Today, the viability of outsourcing as a sustainable business model for the industry has been proven, and FSA is focused on the perpetuation of this business model throughout the worldwide semiconductor industry.

Mission & Vision

FSA's mission is to positively impact the growth and return on invested capital of the global fabless business model to enhance the environment for innovation by providing a platform for meaningful global collaboration between fabless companies and their partners. FSA identifies, debates and discusses business issues in a focused effort to impact solutions and provides members with timely research and resources.

FSA's vision is that the fabless business model is universally recognized as the highest value segment of the semiconductor industry because of its superior innovation, growth and return on invested capital. As this vision becomes a reality, FSA is working to educate the industry and enable fabless companies to gain greater market share and success.

FSA facilitates the relationship between fabless companies and their supplier partners and works to bring the global supply chain together. The diversity in its members enhances the overall leadership and scope of FSA activities.

FSA enhances its global perspectives by bringing even greater value to members to:

- Identify, debate and discuss business and technical issues, providing a focused effort to impact solutions to industry challenges
- Promote the fabless business model worldwide
- Leverage the semiconductor infrastructure/ecosystem
- Provide members with timely global research, resources, publications and survey information.

FSA includes hundreds of corporate members representing a broad cross-section of the industry, including fabless companies and their supply-chain and service partners.

Index

G

H

I

K

L

M

N

O

P

Q

R

S

T